Unit 1

Inspire Science

Earth and Space

Exploring Space

Phenomenon: Why is Saturn different from Earth?

Saturn is the sixth planet from the Sun. It has seven groups of rings and more than 53 moons orbiting it. The strong gravitational pull of this gas giant keeps these objects in orbit.

Scientists think some of Saturn's moons may be habitable!

FRONT COVER: NASA/JPL-Caltech/Space Science Institute. **BACK COVER:** NASA/JPL-Caltech/Space Science Institute.

mheducation.com/prek-12

Send all inquiries to:
McGraw-Hill Education
STEM Learning Solutions Center
8787 Orion Place
Columbus, OH 43240

ISBN: 978-0-07-688282-3
MHID: 0-07-688282-9

Printed in the United States of America.

13 14 15 16 BAB 28 27 26 25 24

McGraw-Hill is committed to providing instructional materials in Science, Technology, Engineering, and Mathematics (STEM) that give all students a solid foundation, one that prepares them for college and careers in the 21st century.

Welcome to

Inspire Science

Earth and Space

Explore Our Phenomenal World

Learning begins with curiosity. Inspire Science is designed to spark your interest and empower you to ask more questions, think more critically, and generate innovative ideas.

Start exploring now!

Inspire Curiosity • Inspire Investigation • Inspire Innovation

Program Authors

Alton L. Biggs
Biggs Educational Consulting
Commerce, TX

Ralph M. Feather, Jr., PhD
Professor of Educational Studies and Secondary Education
Bloomsburg University
Bloomsburg, PA

Douglas Fisher, PhD
Professor of Teacher Education
San Diego State University
San Diego, CA

Page Keeley, MEd
Author, Consultant, Inventor of Page Keeley Science Probes
Maine Mathematics and Science Alliance
Augusta, ME

Michael Manga, PhD
Professor
University of California, Berkeley
Berkeley, CA

Edward P. Ortleb
Science/Safety Consultant
St. Louis, MO

Dinah Zike, MEd
Author, Consultant, Inventor of Foldables®
Dinah Zike Academy, Dinah-Might Adventures, LP
San Antonio, TX

Advisors

Phil Lafontaine
NGSS Education Consultant
Folsom, CA

Donna Markey
NBCT, Vista Unified School District
Vista, CA

Julie Olson
NGSS Consultant
Mitchell Senior High/Second Chance High School
Mitchell, SD

Content Consultants

Chris Anderson
STEM Coach and Engineering Consultant
Cinnaminson, NJ

Emily Miller
EL Consultant
Madison, WI

Key Partners

American Museum of Natural History
The American Museum of Natural History is one of the world's preeminent scientific and cultural institutions. Founded in 1869, the Museum has advanced its global mission to discover, interpret, and disseminate information about human cultures, the natural world, and the universe through a wide-ranging program of scientific research, education, and exhibition.

PhET Interactive Simulations
The PhET Interactive Simulations project at the University of Colorado Boulder provides teachers and students with interactive science and math simulations. Based on extensive education research, PhET simulations engage students through an intuitive, game-like environment where students learn through exploration and discovery.

SpongeLab Interactives
SpongeLab Interactives is a learning technology company that inspires learning and engagement by creating gamified environments that encourage students to interact with digital learning experiences. Students participate in inquiry activities and problem-solving to explore a variety of topics through the use of games, interactives, and video while teachers take advantage of formative, summative, or performance-based assessment information that is gathered through the learning management system.

AdvancED | Measured Progress, a nonprofit organization, is a pioneer in authentic, standards-based assessments with a focus on data-driven tools for improvement. Teachers and students are provided with meaningful assessment that includes robust performance tasks. The assessment content enables teachers to monitor student progress toward learning NGSS.

Table of Contents
Exploring Space

Module 1 The Sun-Earth-Moon System

Encounter the Phenomenon 3

STEM Module Project Launch 4

Lesson 1 Earth's Motion Around the Sun 5
- **Science Probe** Seasons 5
- **Encounter the Phenomenon** 7
- **Explain the Phenomenon** Claim/Evidence/Reasoning Chart 8
- **Investigation** Night and Day 10
- **Investigation** Star Gazing 12
- **Investigation** Ahead of the Curve 13
- **A Closer Look** Polar Night & Midnight Sun 21
- **Review** 22

Lesson 2 Lunar Phases 25
- **Science Probe** Phases of the Moon 25
- **Encounter the Phenomenon** 27
- **Explain the Phenomenon** Claim/Evidence/Reasoning Chart 28
- **Investigation** Foil Moon 30
- **Investigation** The Motion of the Moon 31
- **LAB** Moon Phases 33
- **Science & Society** Return to the Moon 37
- **Review** 38

Lesson 3 Eclipses 41
- **Science Probe** Eclipses 41
- **Encounter the Phenomenon** 43
- **Explain the Phenomenon** Claim/Evidence/Reasoning Chart 44
- **LAB** Beyond a Shadow of a Doubt 46
- **LAB** Casting Shadows 49
- **Investigation** Eclipse Essentials 54
- **A Closer Look** Solar Eclipse Eye Safety 57
- **Review** 58

STEM Module Project Science Challenge: Patterns in the Sky 61

Module Wrap-Up 69

Module 2 Exploring the Universe

Encounter the Phenomenon 71

STEM Module Project Launch 72

Lesson 1 Gravity and the Universe 73

Science Probe Gravity in Space? 73
Encounter the Phenomenon 75
Explain the Phenomenon Claim/Evidence/Reasoning Chart 76
Investigation What Goes Up Must Come Down 78
LAB Changing Shape 80
Careers in Science History from Space 83
LAB Elliptical Orbits 84
Investigation Classification of Galaxies 86
Review 88

Lesson 2 The Solar System 91

Science Probe Objects in Our Solar System 91
Encounter the Phenomenon 93
Explain the Phenomenon Claim/Evidence/Reasoning Chart 94
Investigation Compare the View 96
Investigation Graphing Characteristics 98
LAB Model the Inner Planets 100
LAB Scale Down 101
Investigation Digging Deeper 104
Investigation Moons of the Outer Planets 106
Careers in Science Pluto 109
Review 110

STEM Module Project Science Challenge: Wanted: Space Investigator 113

Module Wrap-Up 119

The Sun-Earth-Moon System

ENCOUNTER
THE PHENOMENON

How do the movements of the Sun, Earth, and the Moon influence the seasons, phases of the Moon, and eclipses?

Sunrise Over Earth

GO ONLINE
Watch the video *Sunrise Over Earth* to see this phenomenon in action.

Communicate Think about seasons, eclipses, and the phases of the Moon. Record your ideas for how or why you think these events happen below. Discuss your ideas with three different partners. Revise or update your ideas, if necessary, after the discussions with your classmates.

STEM Module Project Launch
Science Challenge

Patterns in the Sky

Imagine you are an astronomer who works for a science museum. A group of third-grade students is visiting the museum to learn more about the Sun-Earth-Moon system.

You have been asked to develop a model to help students visualize the patterns of lunar phases, eclipses of the Sun and the Moon, and seasons. To test your model, you will use it to predict different patterns in the sky.

Start Thinking About It

In the image above you see an eclipse of the Sun. How do scientists use models when studying eclipses, lunar phases, and seasons? Why do scientists use models?

STEM Module Project

Planning and Completing the Science Challenge

How will you meet this goal? The concepts you will learn throughout this module will help you plan and complete the Science Challenge. Just follow the prompts at the end of each lesson!

LESSON 1 LAUNCH

Seasons

Six friends wondered why it is hot in the summer and cold in the winter. This is what they said:

Mario: I think Earth is closer to the Sun in the summertime.

Louie: I think the Sun gives off more heat in the summertime.

Brad: I think Earth rotates toward the Sun in summertime.

Chelsea: I think Earth tilts towards the Sun in summer and away from the Sun in winter.

April: I think the sunlight that hits Earth is spread out more in the summertime.

Pedro: I don't think it is any of things you described. I think it is something else.

Which friend do you agree with the most? ______________________ Explain why you agree with that friend.

____

____

____

____

You will revisit your response to the Science Probe at the end of the lesson.

LESSON 1

Earth's Motion Around the Sun

ENCOUNTER THE PHENOMENON | What causes the seasons?

Imagine you and your family go hiking in the same woods every year during each season. At noon you always stop for a picnic lunch at the same clearing at the top of the same hill. You always noticed that the scenery changes from spring to summer to fall to winter, but one year you notice that the Sun's apparent path across the sky changes from season to season. The Sun appears lowest in the sky during the winter and highest in the sky during the summer. Diagram why you think the Sun's position in the sky changes each season in the space below. Then, support your diagram with your reasoning.

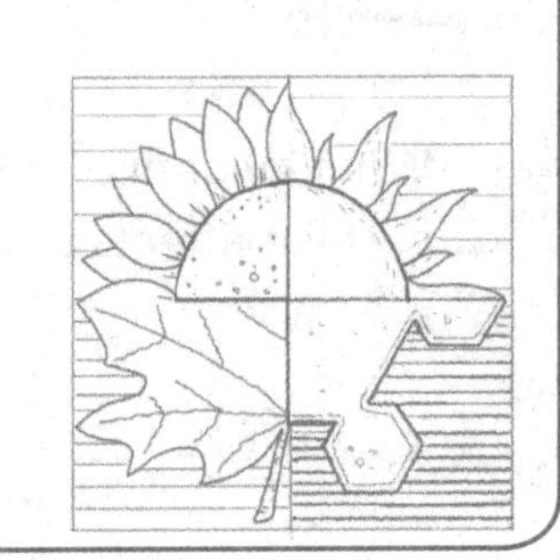

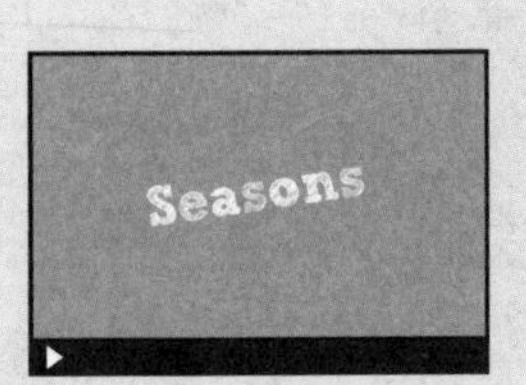

GO ONLINE
Watch the video *Seasons* to see this phenomenon in action.

ENCOUNTER THE PHENOMENON

In the previous activity, you diagrammed why the Sun appears to change location in the sky over the course of a year. How might this phenomena relate to changing seasons? Use your observations about the phenomenon to make a claim about what causes seasons.

CLAIM

Seasons are caused by...

COLLECT EVIDENCE as you work through the lesson. Then return to these pages to record your evidence.

EVIDENCE

A. What evidence have you discovered to explain how Earth moves?

B. What evidence have you discovered to explain how the curve of Earth's surface affects temperatures on Earth?

MORE EVIDENCE

C. What evidence have you discovered to explain the relationship between Earth's tilt and seasons?

When you are finished with the lesson, review your evidence. If necessary, based on the evidence, revise your claim.

REVISED CLAIM

Seasons are caused by...

Finally, explain your reasoning for how and why your evidence supports your claim.

REASONING

The evidence I have collected supports my claim because...

How does Earth move?

If you look outside at the ground, trees, and buildings, it does not seem like Earth is moving. Yet Earth is always in motion, spinning in space and traveling around the Sun. Earth's motion changes how energy from the Sun spreads out over Earth's surface. Let's investigate further.

INVESTIGATION

Night and Day

1. Have you ever thought about the causes of night and day? Observe the demonstration your teacher performs and record your observations, including drawings of what you observe.

2. Do you think every planet in the solar system has night and day? Why or why not?

Want more information?
Go online to read more about Earth's motion around the Sun.

FOLDABLES®
Go to the Foldables® library to make a Foldable® that will help you take notes while reading this lesson.

Earth's Rotation As Earth moves around the Sun, it spins. A spinning motion is called **rotation.** Some spinning objects rotate on a rod or axle. Earth rotates on an imaginary line through its center. The line on which an object rotates is the **rotation axis.**

Suppose you could look down on Earth's North Pole and watch Earth rotate. You would see that Earth rotates on its rotation axis in a counterclockwise direction, from west to east. One complete rotation of Earth takes about 24 hours. This rotation helps produce Earth's cycle of day and night.

Earth's Revolution Earth moves around the Sun in nearly a circular path. The path an object follows as it moves around another object is an **orbit.** The motion of one object around another is called **revolution.** Earth makes one complete revolution around the Sun every 365.24 days.

Why does Earth orbit the Sun? The answer is the Law of Universal Gravitation. This law states that the pull of gravity between two objects depends on the masses of the objects and the distance between them. The more mass either object has, or the closer together they are, the stronger the gravitational pull.

Earth's Tilted Axis Earth's rotation axis is tilted. The tilt of Earth's rotation axis is always in the same direction by the same amount. This means that during half of Earth's orbit, the north end of the rotation axis is tilted toward the Sun. During the other half of Earth's orbit, the north end of the rotation axis is tilted away from the Sun.

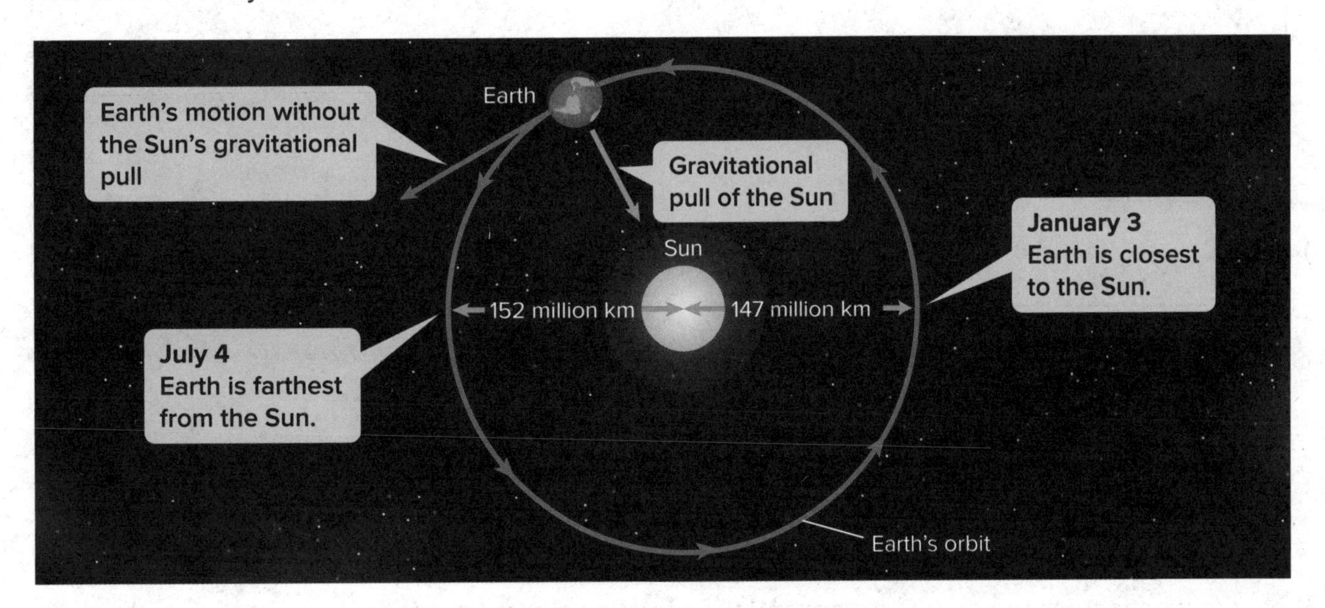

The Sun's effect on Earth's motion is illustrated above. Earth's motion around the Sun is like the motion of an object twirled on a string. The string pulls on the object and makes it move in a circle. If the string breaks, the object flies off in a straight line. In the same way, the pull of the Sun's gravity keeps Earth revolving around the Sun in a nearly circular orbit. If the gravity between Earth and the Sun were to somehow stop, Earth would fly off into space in a straight line.

Why does the view of the sky change over time?

Have you ever looked up at the sky on a clear, dark night and seen stars? If you look at a clear night sky for a long time, what do you notice about the stars?

INVESTIGATION

Star Gazing

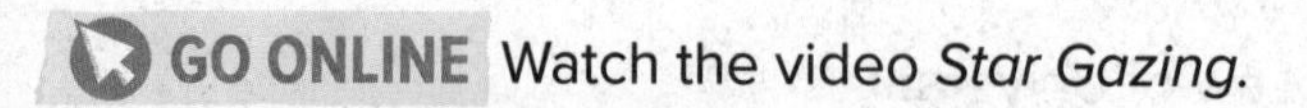

Watch the video *Star Gazing.*

Record your observations about the stars. Describe what you think is happening. Include drawings and illustrations to help explain your observations.

Apparent Motion Each day the Sun appears to move from east to west across the sky. It seems as if the Sun is moving around Earth. However, it is Earth's rotation that causes the Sun's apparent motion.

Earth rotates from west to east. As a result, the Sun appears to move from east to west across the sky. The stars and the Moon also seem to move from east to west across the sky due to Earth's west to east rotation.

To better understand this, imagine riding on a merry-go-round. As you and the ride move, people on the ground appear to be moving in the opposite direction. In the same way, as Earth rotates from west to east, the Sun appears to move from east to west.

COLLECT EVIDENCE

What evidence have you discovered to explain how Earth moves? Record your evidence (A) in the chart at the beginning of the lesson.

Why is Earth warmer at the equator and colder at the poles?

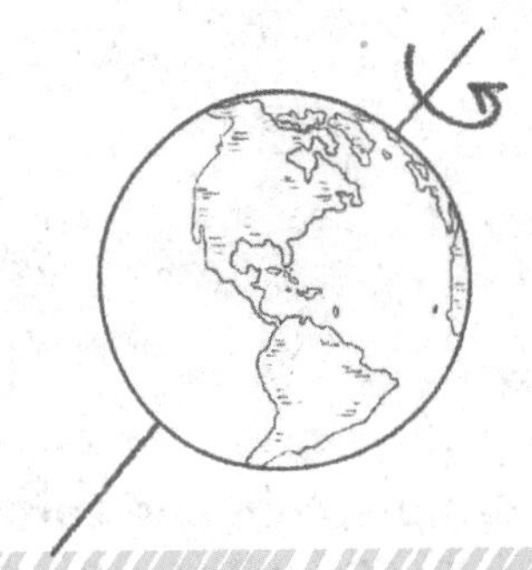

Temperatures near Earth's poles are colder than temperatures near the equator. What causes these temperature differences? Let's find out!

INVESTIGATION

Ahead of the Curve

Observe the demonstration your teacher conducts using a globe and a flashlight. Illustrate and describe what you observe below.

1. Compare and contrast the shapes you drew above.

2. At which location on the globe is the light more spread out? Explain your response.

Temperature and Earth's Curved Surface As Earth orbits the Sun, only one half of Earth faces the Sun at a time. A beam of sunlight carries energy. The more sunlight that reaches a part of Earth's surface, the warmer that part becomes. Because Earth's surface is curved, different parts of Earth's surface receive different amounts of the Sun's energy.

Energy Received by a Tilted Surface Suppose you shine a beam of light on a flat card. As you tilt the card relative to the direction of the light beam, light becomes more spread out on the card's surface. As a result, the energy that the light beam carries also spreads out more over the card's surface. An area on the surface within the light beam receives less energy when the surface is more tilted relative to the light beam.

The Tilt of Earth's Curved Surface Instead of being flat like a card, Earth's surface is curved. Relative to the direction of a beam of sunlight, Earth's surface becomes more tilted as you move north or south from the equator. As shown below, the energy in a beam of sunlight tends to become more spread out the farther you travel north or south from the equator. This means that regions near the poles receive less energy than regions near the equator. This makes Earth colder at the poles and warmer at the equator.

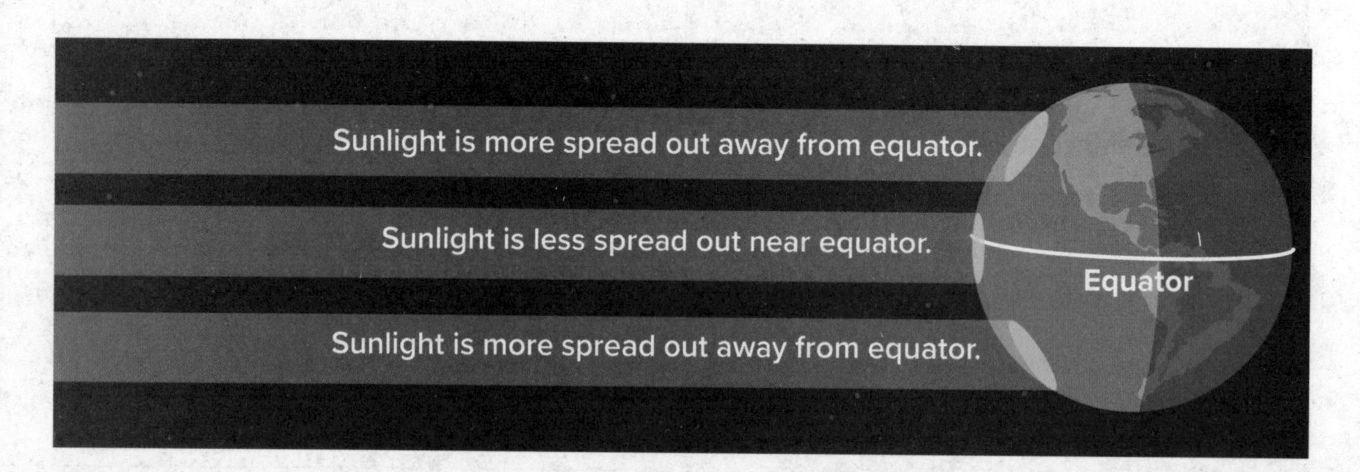

THREE-DIMENSIONAL THINKING

Based on your observations from the *Ahead of the Curve* investigation and the **model** you used, how is the Sun's energy received on Earth? What **effect** does Earth's curved surface have on temperatures?

COLLECT EVIDENCE

What evidence have you discovered to explain how the curve of Earth's surface affects temperatures on Earth? Record your evidence (B) in the chart at the beginning of the lesson.

Why do Earth's seasons change as Earth orbits the Sun?

You might think that summer occurs in the northern hemisphere when Earth is closest to the Sun, and winter occurs in the southern hemisphere when Earth is farthest from the Sun. However, seasonal changes do not depend on Earth's distance from the Sun. In fact, Earth is closest to the Sun in January! Instead, it is the tilt of Earth's rotation axis, combined with Earth's motion around the Sun, that causes the seasons to change.

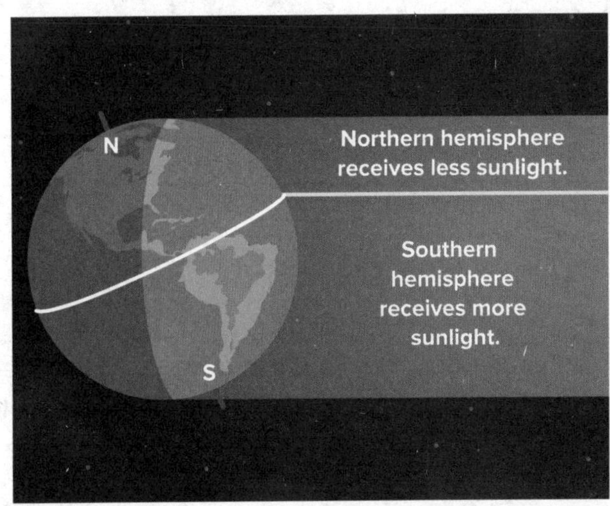

Fall and Winter in the Northern Hemisphere
During one half of Earth's orbit, the north end of the rotation axis is away from the Sun. Then, the northern hemisphere receives less solar energy than the southern hemisphere. Temperatures decrease in the northern hemisphere and increase in the southern hemisphere. This is when fall and winter occur in the northern hemisphere, and spring and summer occur in the southern hemisphere.

Spring and Summer in the Northern Hemisphere During the other half of Earth's orbit, the north end of the rotation axis is toward the Sun. Then, the northern hemisphere receives more energy from the Sun than the southern hemisphere. Temperatures increase in the northern hemisphere and decrease in the southern hemisphere. Days last longer in the northern hemisphere, and nights last longer in the southern hemisphere. This is when spring and summer occur in the northern hemisphere, and fall and winter occur in the southern hemisphere.

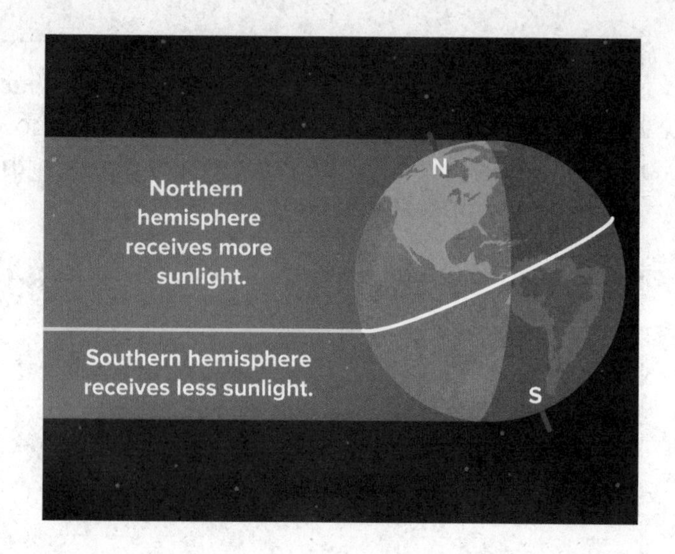

THREE-DIMENSIONAL THINKING

How does the varying amount of the Sun's solar energy **cause** the seasons? What **effects** does the tilt of Earth's rotation axis have on the seasons?

What is Earth's seasonal cycle?

You've learned that as Earth travels around the Sun, its rotational axis always points in the same direction in space. However, the amount that Earth's rotation axis is toward or away from the Sun changes. This causes changes in a yearly cyclical pattern.

There are four days each year when the direction of Earth's rotation axis is special relative to the Sun. A **solstice** is a day when Earth's rotation axis is the most toward or away from the Sun. An **equinox** is a day when Earth's rotation axis is leaning along Earth's orbit, neither toward nor away from the Sun.

March Equinox to June Solstice When the north end of the rotation axis gradually points more and more toward the Sun, the northern hemisphere gradually receives more solar energy. This is spring in the northern hemisphere.

June Solstice to September Equinox The north end of the rotation axis continues to point toward the Sun but does so less and less. The northern hemisphere starts to receive less solar energy. This is summer in the northern hemisphere.

September Equinox to December Solstice The north end of the rotation axis now points more and more away from the Sun. The northern hemisphere receives less and less solar energy. This is fall in the northern hemisphere.

December Solstice to March Equinox The north end of the rotation axis continues to point away from the Sun but does so less and less. The northern hemisphere starts to receive more solar energy. This is winter in the northern hemisphere.

Changes in the Sun's Apparent Path Across the Sky The figure below shows how the Sun's apparent path through the sky changes from season to season in the northern hemisphere. The Sun's apparent path through the sky in the northern hemisphere is lowest on the December solstice and highest on the June solstice.

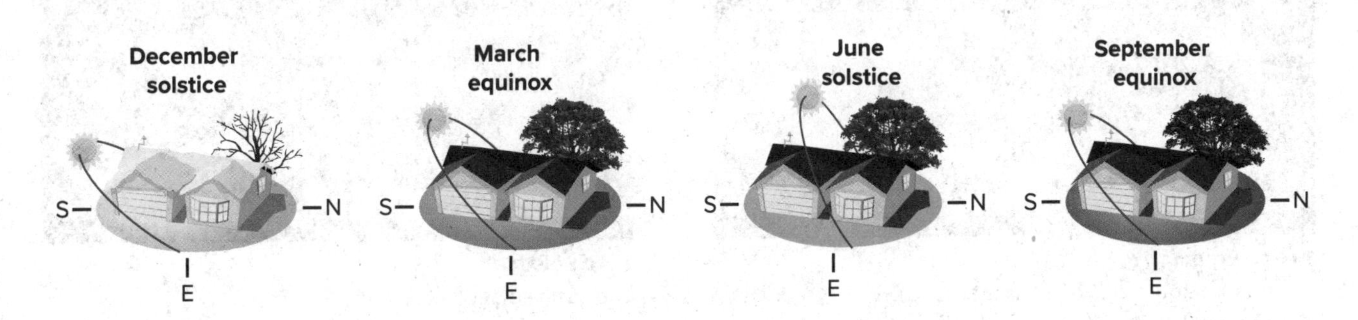

The seasons change as Earth moves around the Sun. Earth's motion around the Sun causes Earth's tilted rotation axis to be leaning toward the Sun and away from the Sun. Let's examine what this looks like.

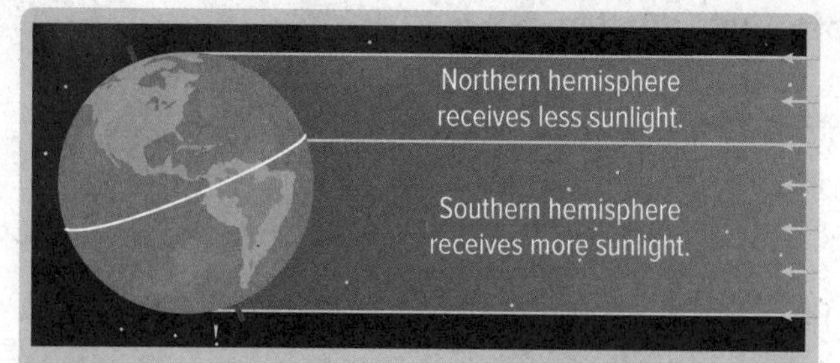

December Solstice
The December solstice is on December 21 or 22.
On this day
- the north end of Earth's rotation axis is away from the Sun;
- days in the northern hemisphere are shortest and nights are longest; winter begins;
- days in the southern hemisphere are longest and nights are shortest; summer begins.

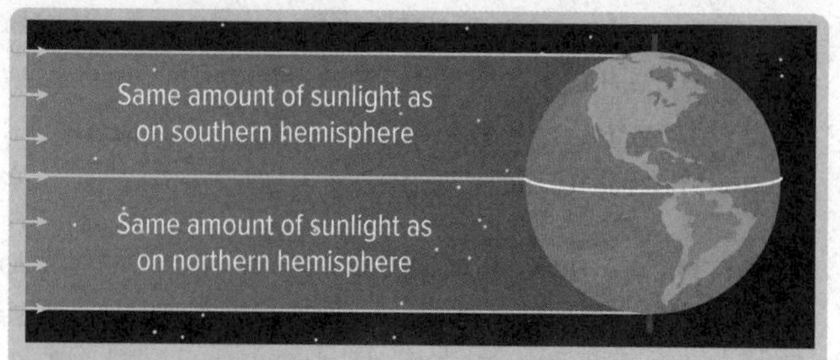

September Equinox
The September equinox is on September 22 or 23.
On this day
- the north end of Earth's rotation axis leans along Earth's orbit;
- there are about 12 hours of daylight and 12 hours of darkness everywhere on Earth;
- autumn begins in the northern hemisphere;
- spring begins in the southern hemisphere.

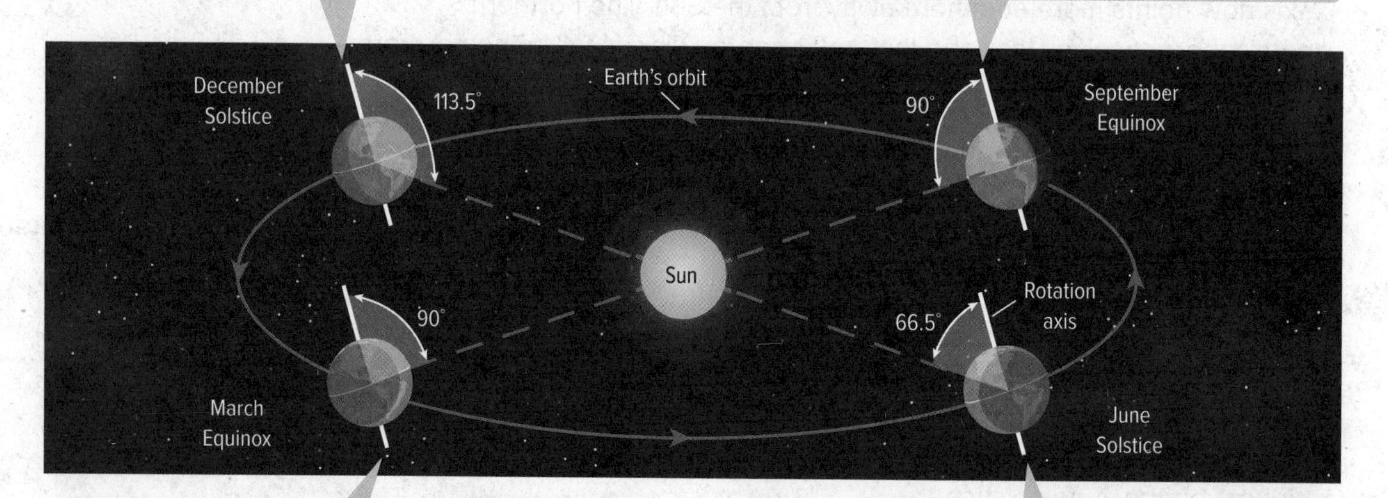

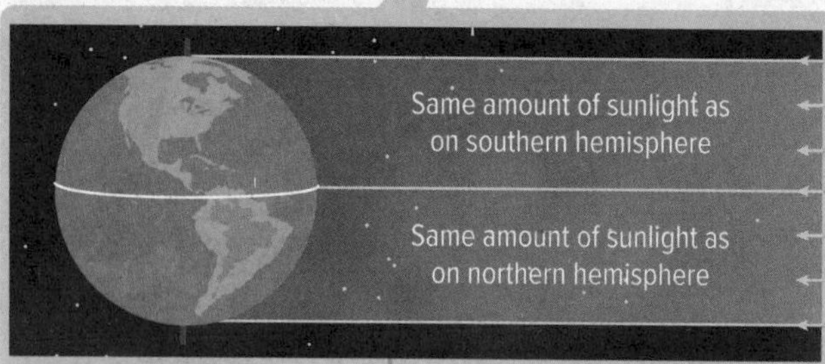

March Equinox
The March equinox is on March 20 or 21.
On this day
- the north end of Earth's rotation axis leans along Earth's orbit;
- there are about 12 hours of daylight and 12 hours of darkness everywhere on Earth;
- spring begins in the northern hemisphere;
- autumn begins in the southern hemisphere.

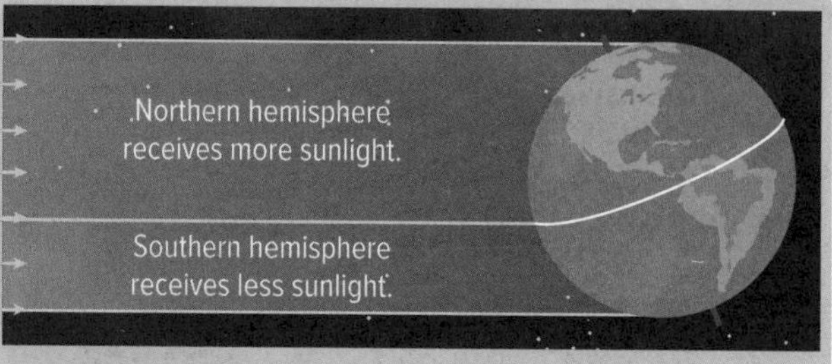

June Solstice
The June solstice is on June 20 or 21.
On this day
- the north end of Earth's rotation axis is toward the Sun;
- days in the northern hemisphere are longest and nights are shortest; summer begins;
- days in the southern hemisphere are shortest and nights are longest; winter begins.

Read a Scientific Text

Astronomical seasons, which are based on the position of Earth in relation to the Sun, differ from the meteorological seasons. Read on to find out why!

CLOSE READING

Inspect

Read the passage *Meteorological Versus Astronomical Seasons*.

Find Evidence

Reread the fourth paragraph. Underline the relationships among events that can be described as cause-and-effect relationships.

Make Connections

Communicate With your partner, discuss how meteorological observing led to the creation of meteorological seasons. How do meteorological seasons differ from astronomical seasons?

PRIMARY SOURCE

Meteorological Versus Astronomical Seasons

You may have noticed that meteorologists and climatologists define seasons differently from "regular" or astronomical spring, summer, fall, and winter. So, why do meteorological and astronomical season begin and end at different times? In short, it's because the astronomical seasons are based on the position of Earth in relation to the sun, whereas the meteorological seasons are based on the annual temperature cycle.

People have used observable periodic natural phenomena to mark time for thousands of years. The natural rotation of Earth around the sun forms the basis for the astronomical calendar, in which we define seasons with two solstices and two equinoxes. Earth's tilt and the sun's alignment over the equator determine both the solstices and equinoxes.

The equinoxes mark the times when the sun passes directly above the equator. In the Northern Hemisphere, the summer solstice falls on or around June 21, the winter solstice on or around December 22, the vernal or spring equinox on or around March 21, and the autumnal equinox on or around September 22. These seasons are reversed but begin on the same dates in the Southern Hemisphere.

Because Earth actually travels around the sun in 365.24 days, an extra day is needed every fourth year, creating what we know as Leap Year. This also causes the exact date of the solstices and equinoxes to vary. Additionally, the elliptical shape of Earth's orbit around the sun causes the lengths of the astronomical seasons to vary between 89 and 93 days. These variations in season length and season start would make it very difficult to consistently compare climatological statistics for a particular season from one year to the next. Thus, the meteorological seasons were born.

Meteorologists and climatologists break the seasons down into groupings of three months based on the annual temperature cycle as well as our calendar. ...

Meteorological observing and forecasting led to the creation of these seasons, and they are more closely tied to our monthly civil calendar than the astronomical seasons are. The length of the meteorological seasons is also more consistent, ranging from 90 days for winter of a non-leap year to 92 days for spring and summer. By following the civil calendar and having less variation in season length and season start, it becomes much easier to calculate seasonal statistics from the monthly statistics, both of which are very useful for agriculture, commerce, and a variety of other purposes.

Source: National Oceanic and Atmospheric Administration

THREE-DIMENSIONAL THINKING

When the northern hemisphere is experiencing summer, Earth is the farthest away from the Sun in its orbit. That distance, about 152,000,000 km, is called the aphelion. Approximately how far is Earth from the Sun in miles? **Solve** the conversion **equation** to calculate the distance in miles. Multiply the distance in km by the conversion factor of 0.62 km. Next find the distance in miles of the perihelion—when Earth is closest to the Sun in its orbit. This is when the northern hemisphere experiences winter.

Aphelion in km: about 152,000,000	Perihelion in km: about 147,000,000
Aphelion in miles:	Perihelion in miles:

If the shape of Earth's orbit was unaltered, but the rotational axis was **changed** so that Earth had no tilt with respect to the orbit, predict the **effect** on the seasons.

COLLECT EVIDENCE

How does the tilt of Earth's rotation axis affect the seasons? Record your evidence (C) in the chart at the beginning of the lesson.

A Closer Look: Polar Night & Midnight Sun

Is it possible to have 24 hours of nighttime or 24 hours of sunlight? Parts of Earth experience these natural phenomena called polar night and midnight sun. How does this happen?

Earth completes one rotation on its tilted axis every 24 hours causing day and night. Because of Earth's tilt, the Sun may never set or may never rise at certain locations on Earth. If the Arctic Circle is tilted toward the Sun, then areas in the Arctic Circle will experience six months of daylight, where the Sun does not set. This begins when the vernal equinox occurs.

While the Arctic Circle experiences midnight sun, the Antarctic Circle will experience polar night. During polar night, the Sun does not rise for six months because the Antarctic Circle is tilted away from the Sun.

Once the autumnal equinox occurs, the opposite phenomenon occurs in each location. The Arctic Circle will experience polar night while the Antarctic Circle will experience midnight sun.

In the northern hemisphere, polar night and midnight sun occur in Norway, parts of Alaska, Canada, Greenland, Finland, Russia, and Sweden. The only place this occurs in the southern hemisphere is in Antarctica.

It's Your Turn

HEALTH Connection How does living six months with 24 hours of sunlight or darkness affect the human body? Research and present your findings to your class.

LESSON 1

Review

Summarize It!

1. **Construct** Diagram how the Sun-Earth system causes the seasonal patterns. Include illustrations and science terms in your diagram.

Three-Dimensional Thinking

2. Which best explains why Earth is colder at the poles than at the equator?

A Earth is farther from the Sun at the poles than at the equator.

B Earth's orbit is not a perfect circle.

C Earth's rotation axis is tilted.

D Earth's surface is more tilted at the poles than at the equator.

Use the figure to answer question 3.

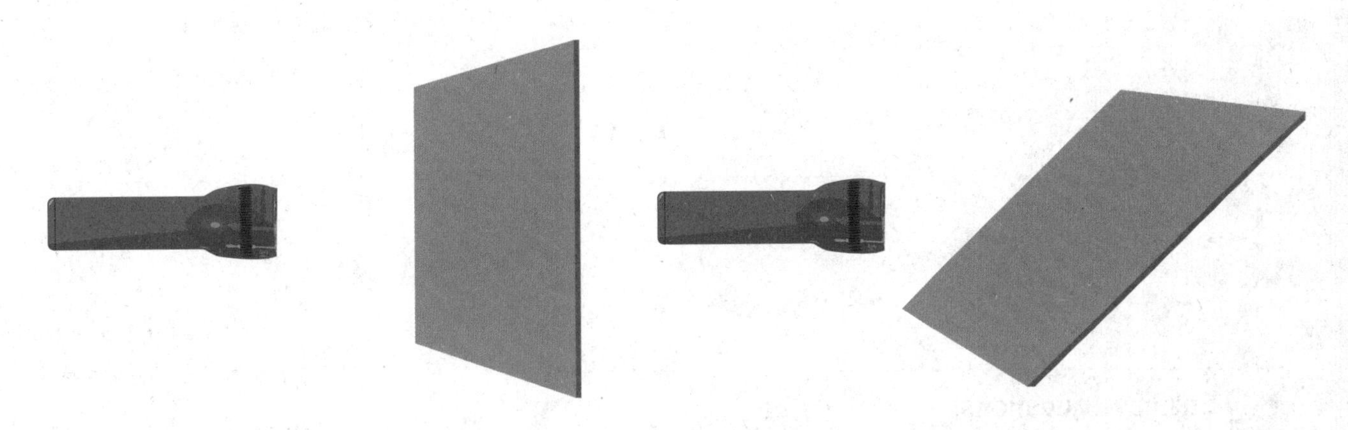

3. Predict the pattern of the light that will appear on each sheet of paper when the flashlight is turned on.

A The pattern on the vertical paper will be a circle, the pattern that will appear on the tilted paper will be elongated and spread out.

B The pattern on the papers will change because the light will fluctuate.

C The pattern that will appear on both pieces of paper is the same—a circle of light.

D The pattern that will appear on both pieces of paper is the same—elongated and spread out.

Real-World Connection

4. **Analyze** How do you think seasonal change affects your local community's usage of energy?

5. **Infer** How do you think animals are affected by seasonal change?

Still have questions?
Go online to check your understanding of Earth's motion and seasons.

REVISIT

Do you still agree with the friend you chose at the beginning of the lesson? Return to the Science Probe at the beginning of the lesson. Explain why you agree or disagree with that friend now.

EXPLAIN THE PHENOMENON

Revisit your claim about what causes the seasonal changes. Explain how your evidence supports your claim.

START PLANNING

STEM Module Project
Science Challenge

Now that you learned about Earth's motion, the tilt of Earth's axis, and the seasons, go to your Module Project to begin incorporating information from this lesson into your model. You will need to be able to show how the seasonal patterns are visible in your model.

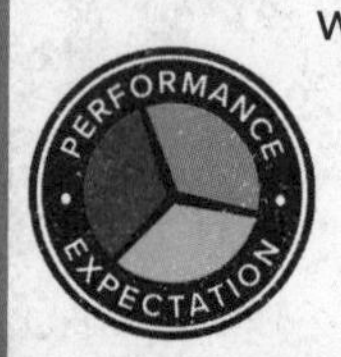

LESSON 2 LAUNCH

PAGE KEELEY SCIENCE PROBES

Phases of the Moon

Many people have different ideas about what causes us to see different parts of the Moon (moon phases). Which idea below best matches your thinking?

A. The Earth casts a shadow on the Moon that allows us to see only the lit part.

B. The Moon moves into the Sun's shadow, blocking out part of the Moon's light.

C. The part we see depends on where the Moon is in relation to Earth and the Sun.

D. Sun's movement around Earth causes different parts of the Moon to be reflected.

E. The Moon's rotation causes different parts of the Moon to be reflected back to Earth.

F. None of these. I think there is something else that causes us to see different moon phases.

Explain your thinking. Describe your ideas about why we see different phases of the Moon.

You will revisit your response to the Science Probe at the end of the lesson.

LESSON 2

Lunar Phases

ENCOUNTER THE PHENOMENON | Why isn't the Moon always full?

Place a ball on a level surface. Position a flashlight so that the light beam shines fully on one side of the ball. Stand behind the flashlight and make a drawing of the ball's appearance in the space below.

Now stand behind the ball and sketch the ball's appearance.

Finally, stand to the left of the ball and illustrate the ball's appearance.

GO ONLINE
Watch the video *Moon Metamorphosis* to see this phenomenon in action.

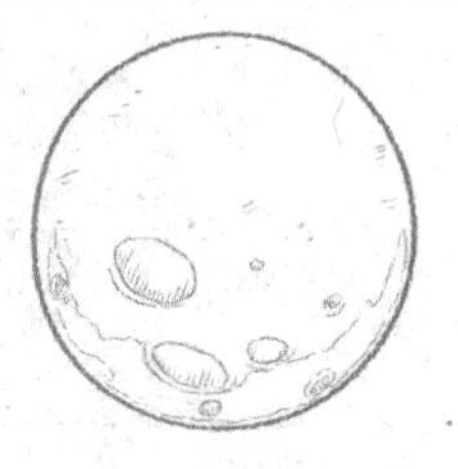

EXPLAIN THE PHENOMENON

You just modeled how the Moon can appear full, only partially lit, or seem to disappear entirely. Are you starting to get some ideas about what might cause this phenomenon? Use your observations from the previous activity to make a claim about the Moon's changing appearance.

CLAIM

The Moon's changing appearance is caused by...

COLLECT EVIDENCE as you work through the lesson. Then return to these pages to record your evidence.

EVIDENCE

A. What evidence have you collected to explain from where the Moon receives its light?

MORE EVIDENCE

B. What evidence have you collected to explain the relationship between the Moon's revolution and lunar phases?

When you are finished with the lesson, review your evidence. If necessary, based on the evidence, revise your claim.

REVISED CLAIM

The Moon's changing appearance is caused by...

Finally, explain your reasoning for how and why your evidence supports your claim.

REASONING

The evidence I collected supports my claim because...

How are we able to see the Moon?

Imagine what people thousands of years ago thought when they looked up at the Moon. They might have wondered why the Moon shines and why it seems to change shape. They probably would have been surprised to learn that the Moon does not emit light at all. Unlike the Sun, the Moon is a solid object that does not emit its own light. So where does the Moon's light come from? Let's investigate!

INVESTIGATION

Foil Moon

1. Observe your teacher's demonstration and write your observations below.

2. Compare the foil-wrapped ball to the unwrapped ball and how the light appeared on each.

3. Make a claim about how these observations might explain where the Moon receives its light.

Light from the Sun The Moon does not produce or emit its own light. The Moon receives its light from the Sun. The Moon's surface is light in color and reflects the light from the Sun.

The amount of light that is reflected off the Moon's surface is actually quite low. Only about 12 percent of the light from the Sun is reflected off the Moon. That 12 percent still makes the Moon the brightest object in our night sky!

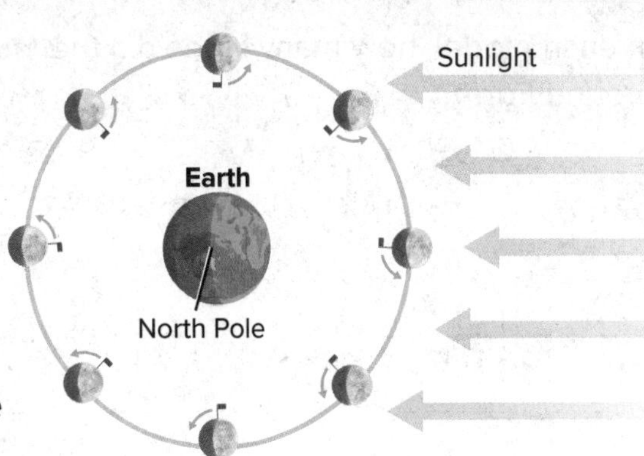

COLLECT EVIDENCE

Where does the Moon receive its light? Record your evidence (A) in the chart at the beginning of the lesson.

How does the Moon move?

While Earth is revolving around the Sun, the Moon is revolving around Earth. The gravitational pull of Earth on the Moon causes the Moon to move in an orbit around Earth. The Moon makes one revolution around Earth every 27.3 days. Does the Moon rotate like Earth? Let's find out!

Want more information?
Go online to read more about the motion of the moon.

INVESTIGATION

The Motion of the Moon

1. Choose a partner. One person represents the Moon. The other represents Earth.
2. For the first model, have the *Moon* move counterclockwise around *Earth*, always facing the same wall. *Earth* should turn counterclockwise to continuously face the *Moon*.
3. For the second model, have the *Moon* move around *Earth* always facing *Earth*. *Earth* should turn counterclockwise to continuously face the *Moon*.
4. For which model was the *Moon* both revolving and rotating?

5. In each model, how many times did the *Moon* rotate during one revolution around *Earth*?

THREE-DIMENSIONAL THINKING

Explain how the Moon can be rotating if the same side of the Moon is always facing Earth.

One Rotation, One Revolution The Moon also rotates as it revolves around Earth. One complete rotation of the Moon also takes 27.3 days. This means the Moon makes one rotation in the same amount of time that it makes one revolution around Earth. Because the Moon makes one rotation for each revolution of Earth, the same side of the Moon always faces Earth. This side of the Moon is called the near side. The side of the Moon that cannot be seen from Earth is called the far side of the Moon.

Why does the Moon appear to change shape?

The Sun is always shining on half of the Moon, just as the Sun is always shining on half of Earth. However, as the Moon moves around Earth, usually only part of the Moon's near side is lit. The portion of the Moon or a planet reflecting light as seen from Earth is called a **phase.** How many phases are there? Let's find out!

FOLDABLES®

Go to the Foldables® library to make a Foldable® that will help you take notes while reading this lesson.

LAB Moon Phases

Safety 

Materials

foam ball

lamp

pencil

stool

Procedure

1. Read and complete a lab safety form.

2. Have one group member hold a foam ball that represents the Moon. Make a handle for the ball by inserting a pencil about two inches into the foam ball. Another group member will represent an observer on Earth. Have the observer sit on a stool and record observations during the activity.

3. Place a lamp on a desk or other flat surface. Remove the shade from the lamp. The lamp represents the Sun.

4. Turn on the lamp and darken the lights in the room.

   CAUTION: Do not touch the bulb or look directly at it after the lamp is turned on.

5. Position the Earth observer's stool about 1 m from the Sun. Position the Moon 0.5–1 m from the observer so that the Sun, Earth, and the Moon are in a line. The student holding the Moon should hold the Moon so it is completely illuminated on one half. The observer records the phase and what the phase looks like in the data table on the next page.

6. Move the Moon counterclockwise about one-eighth of the way around its "orbit" of Earth. The observer swivels on the stool to face the Moon and records the phase.

7. Continue the Moon's orbit until the Earth observer has recorded all the Moon's phases.

8. Return to your positions as the Moon and Earth observer. Choose a part in the Moon's orbit that you did not model. Predict what the Moon would look like in that position, and check if your prediction is correct.

9. Follow your teacher's instructions for proper clean up.

Data and Observations

Position	Phase	Description
1		
2		
3		
4		
5		
6		
7		
8		

Analyze and Conclude

10. Use your observations to explain how the positions of the Sun, the Moon, and Earth produce the different phases of the Moon.

11. Why is half of the Moon always lit? Why do you usually see only part of the Moon's lit half?

12. Based on your observations, why is the Moon not visible from Earth during the new moon phase?

____

____

Lunar Phases While the Moon rotates and revolves around Earth in 27.3 days, the length of the lunar phases cycle is 29.5 days. The Moon's appearance changes as Earth and the Moon move. Depending on where the Moon is in relation to Earth and the Sun, observers on Earth see only part of the light the Moon reflects from the Sun.

Waxing Phases During the **waxing phases,** more of the Moon's near side is lit each night.

Week 1—First Quarter As the lunar cycle begins, a sliver of light can be seen on the Moon's western edge. Gradually the lit part becomes larger. By the end of the first week, the Moon is at its first quarter phase. In this phase, the Moon's entire western half is lit.

First Quarter Moon

Week 2—Full Moon During the second week, more and more of the near side becomes lit. When the Moon's near side is completely lit, it is at the full moon phase.

Full Moon

Waning Phases During the **waning phases,** less of the Moon's near side is lit each night. As seen from Earth, the lit part is now on the Moon's eastern side.

Week 3—Third Quarter During this week, the lit part of the Moon becomes smaller until only the eastern half of the Moon is lit. This is the third quarter phase.

Third Quarter Moon

Week 4—New Moon During this week, less and less of the near side is lit. When the Moon's near side is completely dark, it is at the new moon phase.

New Moon

You've just read that as the Moon revolves around Earth, the part of the Moon's near side that is lit changes. Let's examine what the Moon looks like at different places in its orbit.

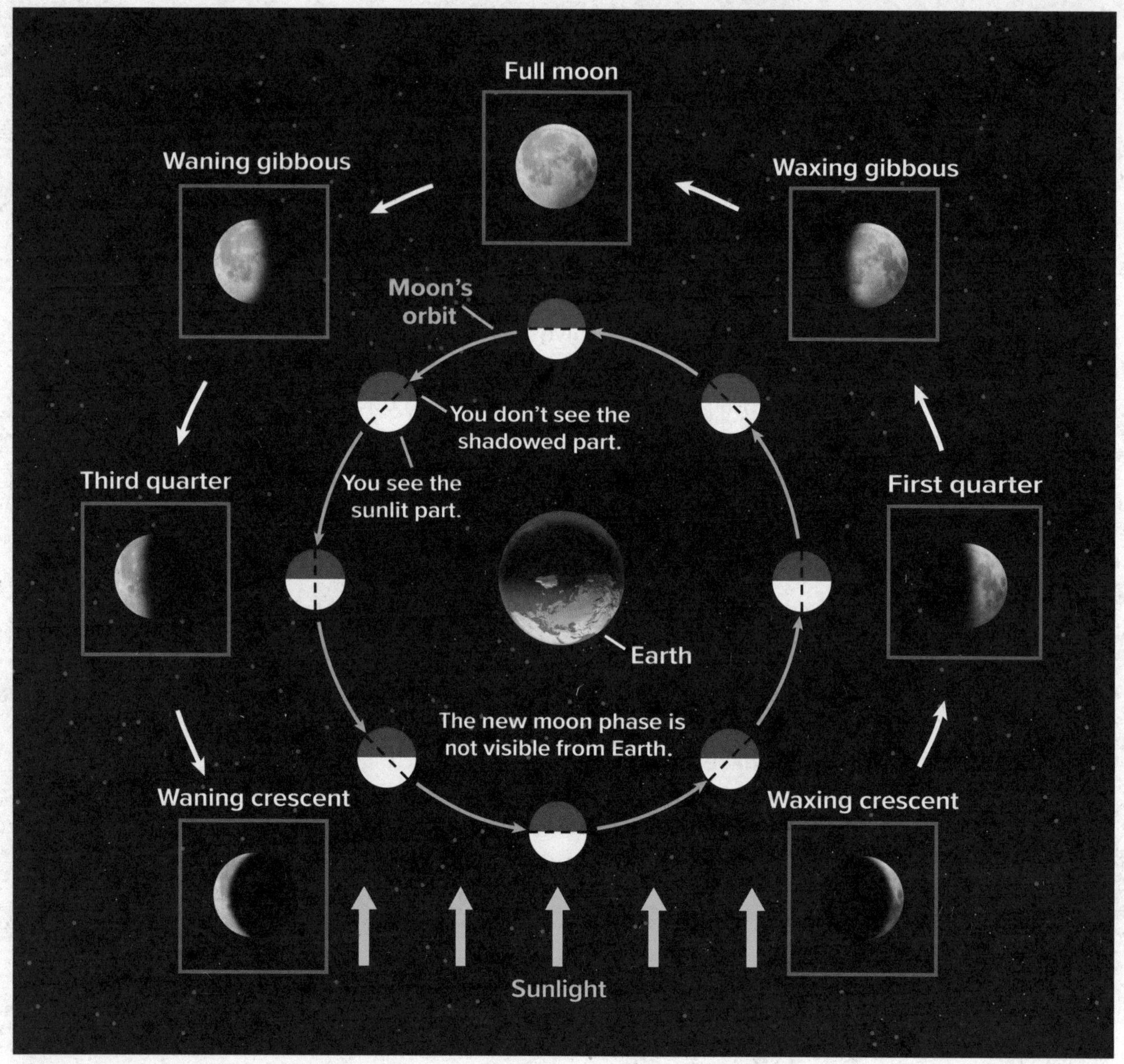

The Moon's motion around Earth causes the Moon to rise, on average, about 50 minutes later each day. The figure below shows how the Moon looks at midnight during three phases of the lunar cycle.

First quarter

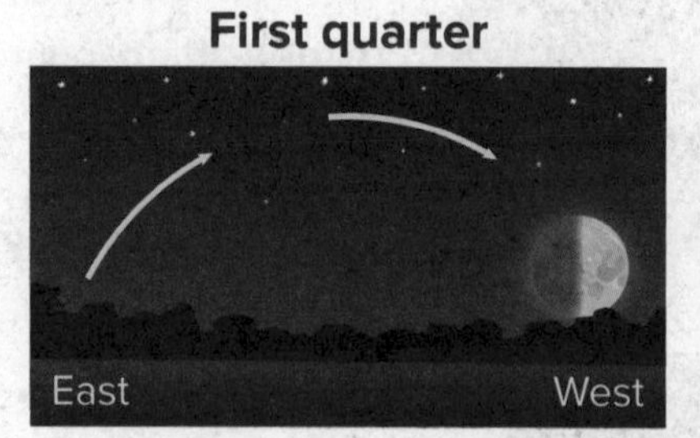

At midnight, the first quarter moon is setting. It rises during the day at about noon.

Full moon

East West

The full moon is highest in the sky at about midnight. It rises at sunset and sets at sunrise.

Third quarter

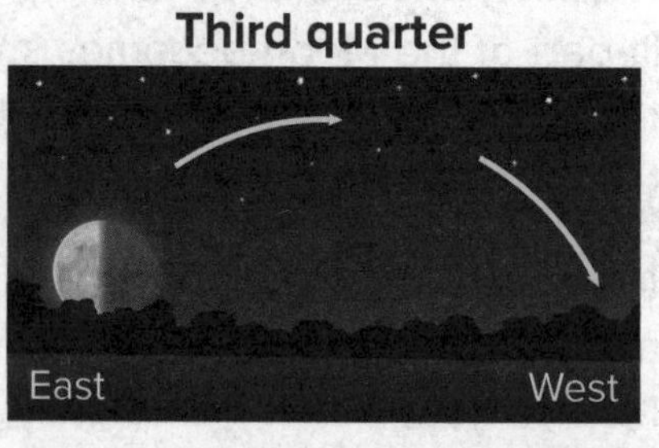

The third quarter moon rises at about midnight, about six hours later than the full moon rises.

COLLECT EVIDENCE

How does the Moon's revolution contribute to lunar phases? Record your evidence (B) in the chart at the beginning of the lesson.

SCIENCE & SOCIETY

Return to the Moon

Exploring Earth's Moon is a step toward exploring other planets and building outposts in space.

The United States undertook a series of human spaceflight missions from 1961–1975 called the Apollo program. The goal of the program was to land humans on the Moon and bring them safely back to Earth. Six of the missions reached this goal. The Apollo program was a huge success, but it was just the beginning.

NASA began another space program that had a goal to return astronauts to the Moon to live and work. However, before that could happen, scientists needed to know more about conditions on the Moon and what materials are available there. Collecting data was the first step. In 2009, NASA launched the Lunar Reconnaissance Orbiter (LRO) spacecraft. The LRO is in an elliptical orbit that takes it over both of the Moon's poles. As it orbits, it collects detailed data that scientists can use to make maps of the Moon's features and resources, such as deep craters that formed on the Moon when comets and asteroids slammed into it billions of years ago. Some scientists predicted that these deep craters contain frozen water.

One of the instruments launched with the LRO was the Lunar Crater Observation and Sensing Satellite (LCROSS). LCROSS observations confirmed the scientists' predictions that water exists on the Moon. A rocket launched from LCROSS impacted the Cabeus crater near the Moon's south pole. The material that was ejected after the rocket's impact included water. NASA's goal of returning astronauts to the Moon was delayed, and their missions now focus on exploring Mars instead. But the discoveries made on the Moon will help scientists develop future missions that could take humans farther into the solar system.

Apollo SPACE PROGRAM

The Apollo Space Program included 17 missions. Here are some milestones:

May 25 1961
President John F. Kennedy announces the goal of landing a man on the Moon by the end of the decade.

December 21–27 1968
Apollo 8 First manned spacecraft orbits the Moon.

July 16–24 1969
Apollo 11 First humans, Neil Armstrong and Buzz Aldrin, walk on the Moon.

July 1971
Apollo 15 Astronauts drive the first rover on the Moon.

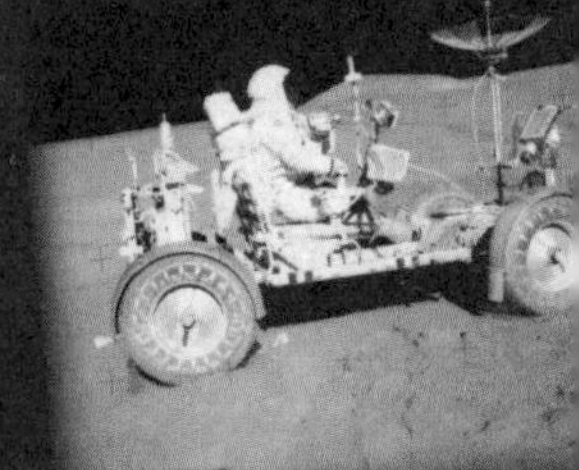

December 7–19 1972
Apollo 17 The first phase of the Moon ended with the last lunar landing mission.

It's Your Turn

Write a paragraph about whether or not you think humans should establish an outpost on the Moon and the importance of water to this effort.

LESSON 2

Review

Summarize it!

1. **Identify** the moon phase represented by each illustration. Then draw what each phase looks like from Earth.

 = Sun  = Moon  = Earth

	Phase		Phase
	Phase		Phase
	Phase		Phase
	Phase		Phase

Three-Dimensional Thinking

2. A new moon occurs once every 29.5 days. Why must the Sun, Earth, and the Moon be aligned in order for the new moon to occur?

A No sunlight is reflected off Earth at this point.

B Sunlight directed toward Earth is blocked by the Moon.

C The Moon does not orbit in the same plane as Earth.

D The Moon is not directing any light toward Earth at this point.

Use the image to answer questions 3 and 4.

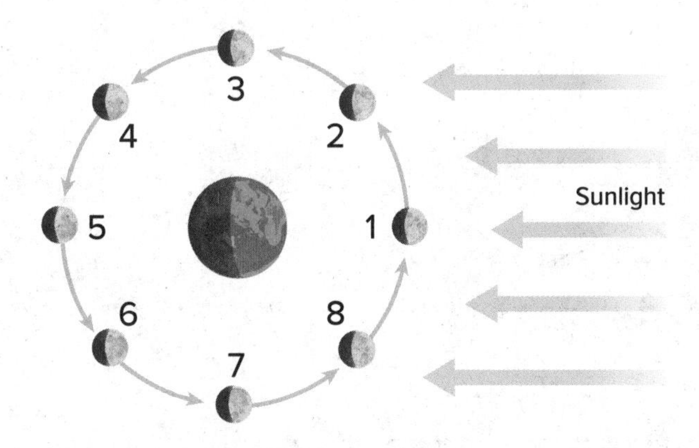

3. Predict the locations when the Moon is in a waxing phase.

A 1, 5

B 2, 3, 4

C 7, 8

D 3, 7

4. Predict which location is the phase seen from Earth at the end of the second week of the lunar cycle.

A 1

B 3

C 5

D 7

Real-World Connection

5. Evaluate Would the phases of the Moon be affected if the Moon did not make one rotation for each revolution of Earth? Explain.

6. Infer Imagine the Moon rotates twice in the same amount of time the Moon orbits Earth once. Would you be able to see the Moon's far side from Earth? Explain your reasoning.

Still have questions?
Go online to check your understanding of lunar phases.

REVISIT

Do you still agree with the statement you chose at the beginning of the lesson? Return to the Science Probe at the beginning of the lesson. Explain why you agree or disagree with that statement now.

EXPLAIN THE PHENOMENON

Revisit your claim about what causes the lunar phases. Explain how your evidence supports your claim.

KEEP PLANNING

STEM Module Project Science Challenge

Now that you learned about the lunar phases, go to your Module Project to continue planning your model. Keep in mind that you want to use your model to help explain the patterns of the lunar phases.

Eclipses

Six students were discussing what causes eclipses. This is what they said:

Isabella: I think an eclipse of the Moon happens about every 29 days.

Sam: I think the Sun stops giving off light during an eclipse.

June: I think we can't see the Moon when Earth's shadow falls on it.

Olivia: I think we can't see the Sun when Earth's shadow falls on it.

Santiago: I think each new moon is the same as an eclipse.

Mateo: I think something else causes eclipses to occur.

Which student do you agree with the most? Explain why you agree with that student.

You will revisit your response to the Science Probe at the end of the lesson.

LESSON 3

Eclipses

ENCOUNTER THE PHENOMENON | What is causing these “bites” out of the Sun?

GO ONLINE Watch the video *Eclipses* to see this phenomenon in action.

Record your observations about the phenomena in the space provided. Why do you think these phenomena occur?

EXPLAIN THE PHENOMENON

You just watched eclipses of the Sun and Moon. Are you starting to get some ideas about what is happening in the Sun-Earth-Moon system during each type of eclipse? Use your observations about the phenomena to make a claim about how eclipses occur.

CLAIM

Eclipses occur because...

COLLECT EVIDENCE as you work through the lesson. Then return to these pages to record your evidence.

EVIDENCE

A. What evidence have you discovered to explain how solar eclipses occur?

B. What evidence have you discovered to explain how lunar eclipses occur?

MORE EVIDENCE

C. What evidence have you discovered to explain the impact of the Moon's tilted orbit on eclipse formation?

When you are finished with the lesson, review your evidence. If necessary, based on the evidence, revise your claim.

REVISED CLAIM

Eclipses occur because...

Finally, explain your reasoning for how and why your evidence supports your claim.

REASONING

The evidence I collected supports my claim because...

What makes a shadow?

A shadow results when one object blocks the light that another object emits or reflects. For example, when a tree blocks light from the Sun, it casts a shadow. If you want to stand in the shadow of a tree, the tree must be in a line between you and the Sun.

Each object in the solar system creates a shadow as it blocks the path of the Sun's light. What happens to an object's shadow when the object moves? Let's find out!

Want more information?
Go online to read more about solar and lunar eclipses.

FOLDABLES®
Go to the Foldables® library to make a Foldable® that will help you take notes while reading this lesson.

LAB Beyond a Shadow of a Doubt

Safety  

Materials

flashlight

opaque object

Procedure

1. Read and complete a lab safety form.
2. Select an object provided by your teacher.
3. Shine a flashlight on the object, projecting its shadow on the wall.
4. While holding the flashlight in the same position, move the object closer to the wall—away from the light. Then, move the object toward the light. Record your observations in the Data and Observations section on the next page.
5. Follow your teacher's instructions for proper cleanup.

Data and Observations

Analyze and Conclude

6. Compare and contrast the shadows created in each situation. Describe the shadows and how they changed.

7. Imagine you could look at the flashlight from behind your object, looking from the darkest and lightest parts of the object's shadow. How much of the light source do you think you could see from each location?

The Umbra and the Penumbra If you go outside on a sunny day and look carefully at a shadow on the ground, you might notice that the edges of the shadow are not as dark as the rest of the shadow. Light from the Sun and other sources casts shadows with two distinct parts, as shown in the figure below.

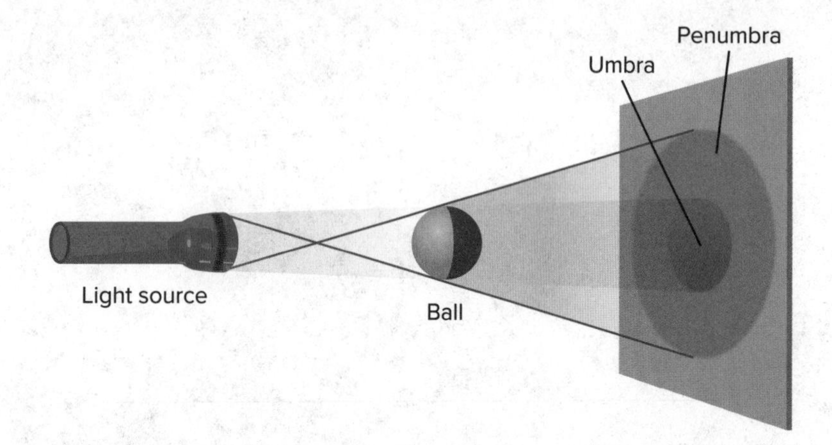

- The **umbra** is the central, darker part of a shadow where light is totally blocked. If you stood within an object's umbra, you would not see the light source at all.
- The **penumbra** is the lighter part of a shadow where light is partially blocked. If you stood within an object's penumbra, you would be able to see only part of the light source.

What is a solar eclipse?

As the Sun shines on the Moon, the Moon casts a shadow that extends out into space. Sometimes the Moon passes between Earth and the Sun. This can only happen during the new moon phase. Let's investigate!

LAB Casting Shadows

Safety 

Materials

flashlight

large foam ball

small foam ball

pencil

Procedure

1. Read and complete a lab safety form.
2. Working with another student, use a pencil to connect two foam balls.
3. While one person holds the foam balls, the other should stand 1 m away and shine a light on the foam balls. The foam balls and light should be in a direct line, with the smallest foam ball closest to the light.
4. Sketch and describe your observations in the Data and Observations section.
5. Follow your teacher's instructions for proper cleanup.

Data and Observations

Analyze and Conclude

6. Explain the relationship between the two types of shadows and solar eclipses.

7. In the lab you used two different-sized foam balls. Which did each of them represent? Now that you know what a solar eclipse is, explain what a lunar eclipse is using words and illustrations.

8. Using what you have learned in this lab, predict how a lunar eclipse occurs.

Solar Eclipse As the Sun shines on the Moon, the Moon casts a shadow that extends out into space. Sometimes the Moon passes between Earth and the Sun. Solar eclipses can only occur during the new moon phase. When Earth, the Moon, and the Sun are lined up, the Moon casts a shadow on Earth's surface. When the Moon's shadow appears on Earth's surface, a **solar eclipse** is occurring. As Earth rotates, the Moon's shadow moves along Earth's surface. The type of eclipse you see—total or partial—depends on where you are in the path of the eclipse.

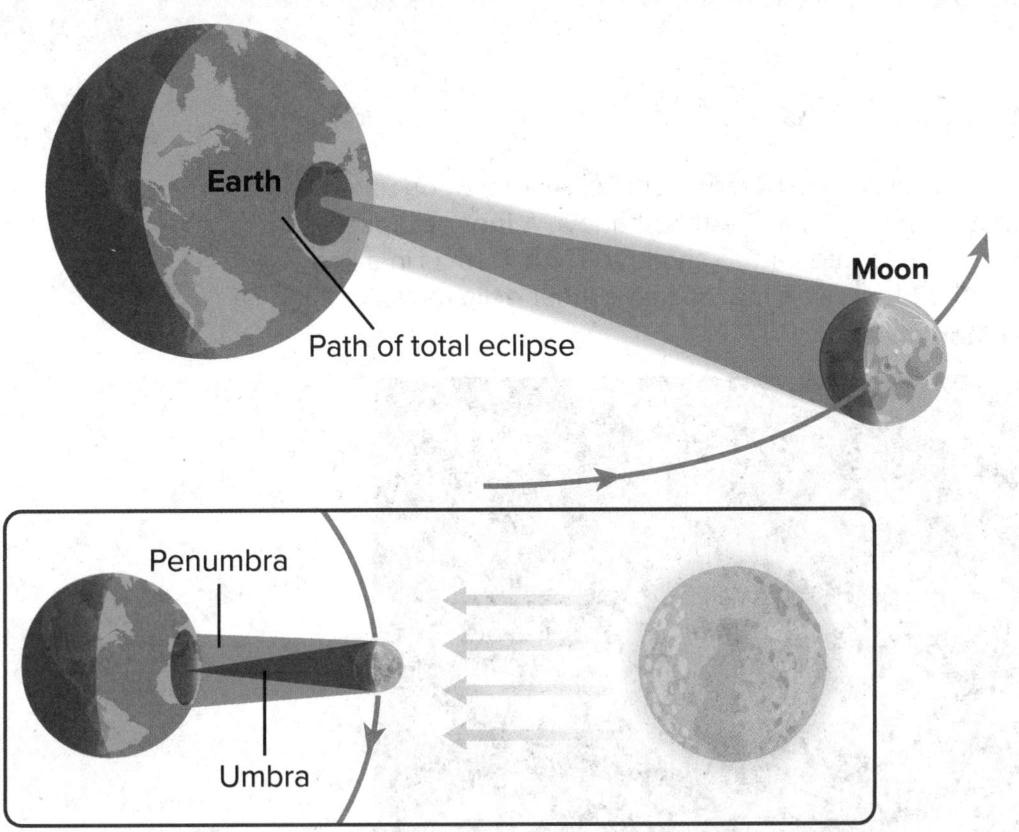

Total Solar Eclipses You can only see a total solar eclipse from within the Moon's umbra. During a total solar eclipse, the Moon appears to cover the Sun completely, as shown in the figure below. Then, the sky becomes dark enough that you can see stars. A total solar eclipse lasts no longer than about seven minutes.

The Sun's Changing Appearance During a Total Solar Eclipse

The Motion of the Moon in the Sky During a Total Solar Eclipse

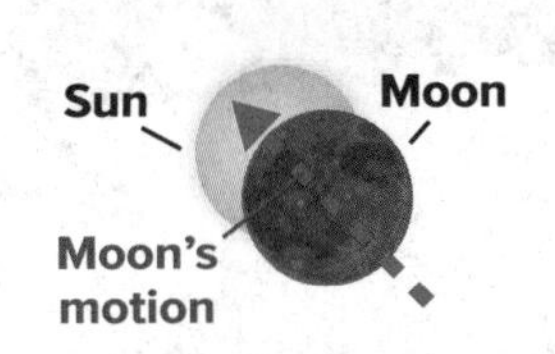

Partial Solar Eclipses You can see a partial solar eclipse from within the Moon's much larger penumbra. The stages of a partial solar eclipse are similar to the stages of a total solar eclipse, except that the Moon never completely covers the Sun.

COLLECT EVIDENCE

What is necessary for a solar eclipse to take place? Record your evidence (A) in the chart at the beginning of the lesson.

What is a lunar eclipse?

Just like the Moon, Earth casts a shadow into space. As the Moon revolves around Earth, it sometimes moves into Earth's shadow. A **lunar eclipse** occurs when the Moon moves into Earth's shadow. Then Earth is in a line between the Sun and the Moon. This means that a lunar eclipse can occur only during the full moon phase.

Like the Moon's shadow, Earth's shadow has an umbra and a penumbra. Different types of lunar eclipses occur depending on which part of Earth's shadow the Moon moves through. Unlike solar eclipses, you can see any lunar eclipse from any location on the side of Earth facing the Moon.

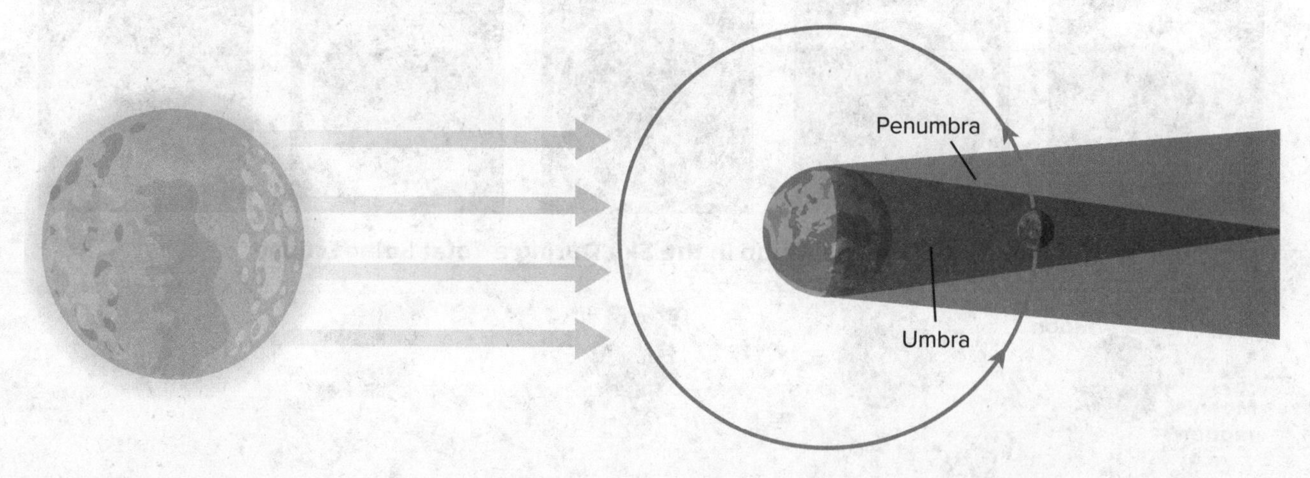

Total Lunar Eclipses When the entire Moon moves through Earth's umbra, a total lunar eclipse occurs. The Moon's appearance changes as it gradually moves into Earth's penumbra, then into Earth's umbra, back into Earth's penumbra, and then out of Earth's shadow entirely.

THREE-DIMENSIONAL THINKING

You've just learned that during a total lunar eclipse, the entire Moon passes through Earth's umbra. The Moon gradually darkens until a dark shadow covers it completely. **Construct an explanation** to describe how a total lunar eclipse would look compared with a total solar eclipse.

____

____

____

You can still see the Moon even within Earth's umbra. Although Earth blocks most of the Sun's rays, Earth's atmosphere deflects some sunlight into Earth's umbra. This is also why you can often see the unlit portion of the Moon on a clear night. This reflected light has a reddish color and gives the Moon a reddish tint during a total lunar eclipse.

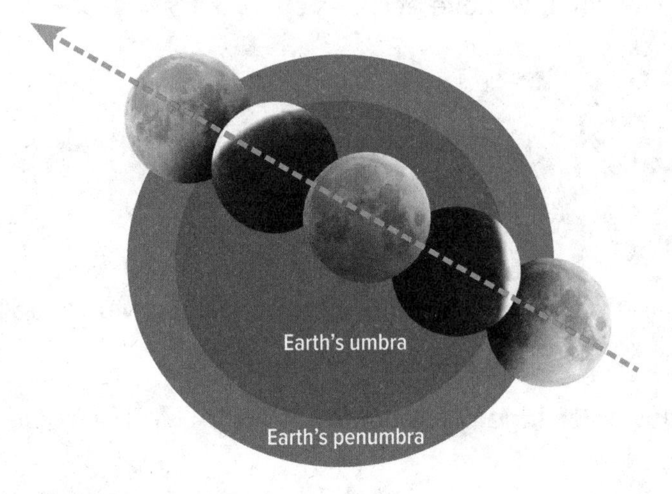

Partial Lunar Eclipses When only part of the Moon passes through Earth's umbra, a partial lunar eclipse occurs. The stages of a partial lunar eclipse are similar to those of a total lunar eclipse, except the Moon is never completely covered by Earth's umbra. The part of the Moon in Earth's penumbra appears only slightly darker, while the part of the Moon in Earth's umbra appears much darker.

COLLECT EVIDENCE

What is necessary for a lunar eclipse to take place? Record your evidence (B) in the chart at the beginning of the lesson.

THREE-DIMENSIONAL THINKING

What **patterns** did you notice between solar and lunar eclipses? How do those patterns impact who can see each type of eclipse? **Explain** your reasoning.

Why aren't there eclipses every month?

A solar eclipse can only happen during the new moon phase. Lunar eclipses can only occur during a full moon. You have learned that each of those phases occurs every 29.5 days. Why don't we have eclipses every 29.5 days? Let's find out!

INVESTIGATION

Eclipse Essentials

GO ONLINE Watch the animation *To Eclipse or Not to Eclipse* and answer the following questions.

1. What causes the Moon's shadow to change placement?

2. From the video, what must happen for a solar eclipse to occur?

3. Why don't we see lunar eclipses every month?

Rare Eclipses Lunar eclipses can only occur during a full moon phase, when the Moon and the Sun are on opposite sides of Earth. However, lunar eclipses do not occur during every full moon because of the tilt of the Moon's orbit with respect to Earth's orbit. During most full moons, the Moon is slightly above or slightly below Earth's penumbra.

Solar eclipses can occur only during a new moon, when Earth and the Sun are on opposite sides of the Moon. However, solar eclipses do not occur during every new moon phase. The figure below shows why. The Moon's orbit is tilted slightly compared to Earth's orbit. As a result, during most new moons, Earth is either above or below the Moon's shadow. However, every so often the Moon is in a line between the Sun and Earth. Then the Moon's shadow passes over Earth and a solar eclipse occurs.

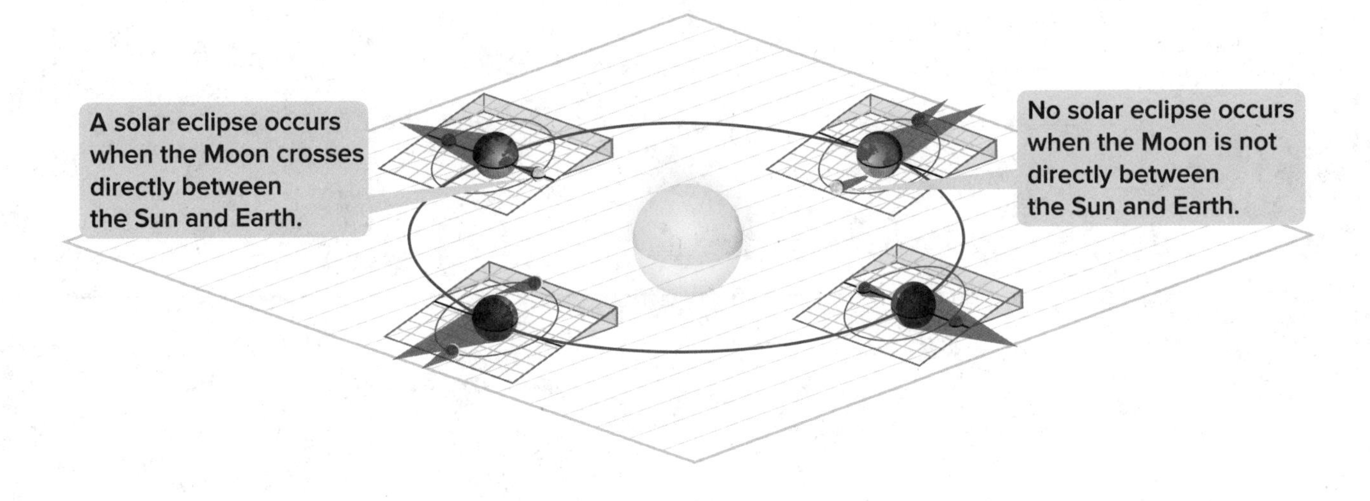

COLLECT EVIDENCE

How does the Moon's tilted orbit affect eclipses? Record your evidence (C) in the chart at the beginning of the lesson.

SEP DCI CCC

THREE-DIMENSIONAL THINKING

You've learned that lunar and solar eclipses do not occur every month. But scientists have been able to detect **patterns** in the Sun–Earth–Moon system and can predict when the next eclipses will occur. Locate an eclipse calendar on the Internet. After examining the calendar and locations, **construct an explanation** of why most people will not witness a total solar eclipse in their lifetime.

What is the next eclipse you may be able to witness? When will the eclipse occur?

GO ONLINE for an additional opportunity to explore!

PHYSICAL SCIENCE Connection Want to know more about eclipses? Investigate eclipses and their applications by performing the following activity.

☐ **Investigate** the relationship between the theory of relativity and solar eclipses in the **Scientific Text** *Testing Einstein's Theory of Gravity.*

A Closer Look: Solar Eclipse Eye Safety

The closest star to Earth is the Sun. The Sun is a star you can observe during the day. Looking directly at the Sun, even for a moment, can be very harmful to your eyes. But is it safe to look at the Sun during a solar eclipse while the Moon blocks the sunlight?

Although the Sun's visible light is blocked by the Moon during a solar eclipse, ultraviolet (UV) rays are still being emitted and could cause damage to your eyes. If you want to observe a solar eclipse, special precautions should be used to stay safe.

To safely observe a solar eclipse, you should use protective eyewear, such as eclipse glasses, telescope filters, or welder's glasses. These items have a thin layer of aluminum, chromium, or silver on the surface that reduces the effects of UV, visible, and infrared light.

It's Your Turn

Research and Present Go online to research when the next solar eclipse will occur. Create a blog to promote safety awareness for viewing a solar eclipse properly.

LESSON 3

Review

Summarize It!

1. Illustrate the positions of the Sun, Earth, and the Moon during a solar eclipse and during a lunar eclipse. Also identify the correct phase that the Moon is in during each type of eclipse.

Solar eclipse

Lunar eclipse

Three-Dimensional Thinking

Use the figure to answer questions 2 and 3.

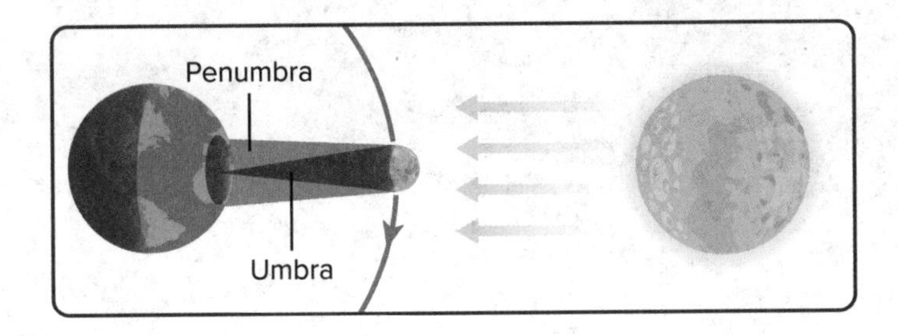

2. What is occurring in this figure?

A lunar eclipse

B lunar and solar eclipse

C partial lunar eclipse

D solar eclipse

3. Why would someone in North America not be able to view this eclipse?

A Because this is a lunar eclipse and Earth casts a very small shadow.

B Because this is a lunar eclipse and North America is experiencing day.

C Because this is a solar eclipse and North America is experiencing night.

D Because this is a solar eclipse and the Moon casts a very small shadow.

---

4. Where do you have to be located to be able to see a total eclipse?

A anywhere on the continent where the eclipse is occuring

B within the penumbra

C within the penumbra and umbra

D within the umbra

Real-World Connection

5. Infer During a lunar eclipse, Earth casts a shadow on the Moon. What type of shadow would Earth cast if it were flat? What type of shadow does Earth cast on the Moon during a lunar eclipse?

6. Predict Many communities rely on solar energy for daily life uses. Suppose the Moon was twice its size and cast a bigger shadow on Earth during a solar eclipse, causing the light from the Sun to be blocked for days. How would this affect the communities who rely on solar energy?

Still have questions?
Go online to check your understanding of solar and lunar eclipses.

REVISIT SCIENCE PROBES Do you still agree with the student you chose at the beginning of the lesson? Return to the Science Probe at the beginning of the lesson. Explain why you agree or disagree with that student now.

EXPLAIN THE PHENOMENON Revisit your claim about what causes eclipses. Explain how your evidence supports your claim.

PLAN AND PRESENT

STEM Module Project
Science Challenge

Now that you learned about the eclipses, go to your Module Project to finish your model. Keep in mind that your model needs to correctly represent the positions of the Sun, Earth, and the Moon when describing eclipses.

STEM Module Project
Science Challenge

Patterns in the Sky

Imagine you are an astronomer who works for a science museum. A group of third-grade students is visiting the museum to learn more about the Sun–Earth–Moon system.

You have been asked to develop a model to help students visualize the patterns of lunar phases, eclipses of the Sun and the Moon, and seasons. To test your model, you will use it to predict different patterns in the sky.

Planning After Lesson 1

In your model, how will you show the distance between Earth and the Sun? How will you show the position of Earth's axis as it orbits the Sun?

Research the different ways you can build and present your models.

Planning After Lesson 1, continued

In the space below, sketch a model showing why seasons occur. Label all components of your model, and include captions explaining Earth's movements and how they are related to seasons. Also describe how sunlight travels and hits different parts of Earth during each season.

Planning After Lesson 2

In your model, how will you show the movements of the Moon and the Earth–Moon system?

Sketch your model of lunar phases in the space below. Label all components of your model, and include captions explaining why the pattern of lunar phases occurs.

PERFORMANCE EXPECTATION

STEM Module Project
Science Challenge

Planning After Lesson 3

In your model, how will you show why solar and lunar eclipses do not occur every month?

Sketch your model of a solar eclipse and a lunar eclipse in the space below. Label all components of your model, and include captions explaining why eclipses occur.

Develop Your Model

Look over the materials provided by your teacher. Then, as a group, design your model. List the materials you will use and each step you will take to build your model. Describe how you will address proportional relationships (size and distance) in your model.

Evaluate Your Model

Once your model is complete, make and record observations of seasons, lunar phases, and solar and lunar eclipses. Then, identify the model elements in the table below.

Model Elements	Descriptions
Components (What are the different parts of my model?)	
Relationships (How do the components of my model interact?)	
Connections (How does my model help me understand the phenomenon?)	

Make Predictions

Use your model to make predictions based on the descriptions below.

Description	Prediction
The Sun, Earth, and the Moon are lined up in a straight row with the Moon in the middle. Predict the phase of the Moon.	
The moon is in the last quarter phase. Predict the relative positions of Earth, the Sun, and the Moon.	
The Sun, Earth, and the Moon form a right angle. Predict whether an eclipse will occur. Explain your prediction.	
A solar eclipse is occurring. Predict the relative positions of the Sun, Earth, and the Moon.	
Earth's southern axis is tilted toward the Sun. Predict the seasons in the southern and northern hemispheres.	
It is summer in the northern hemisphere. Predict the position of Earth in relation to the Sun.	

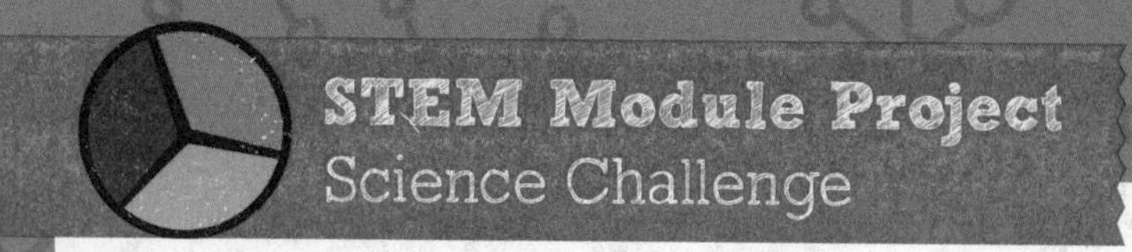

Present Your Model

Present your model before the class.

Is your model appropriate for a third-grade audience? If yes, explain why. If no, explain how you could make it more appropriate.

Why is it difficult to show scale size or scale distance in a model of the Sun-Earth-Moon system?

In addition to problems with scale, what are some limitations of models? What are some advantages?

Congratulations! You've completed the Science Challenge requirements.

Module Wrap-Up

REVISIT THE PHENOMENON

Using the model you developed for the Science Challenge, explain how the Sun, Earth, and the Moon cause seasons, eclipses, and Moon phases.

OPEN INQUIRY

If you had to ask one question about what you studied, what would it be?

Plan and conduct an investigation to answer this question.

Exploring the Universe

ENCOUNTER THE PHENOMENON

What makes up the universe, and how does gravity affect the universe?

Viewing the Universe

GO ONLINE
Watch the video *Viewing the Universe* to see this phenomenon in action.

Collaborate A system is an organized group of components, or objects, that together form a whole. The universe is a system. With your class, develop a list of components that make up this system. Are there any relationships among the components? How does gravity impact the components? Record your thoughts in the space below.

STEM Module Project Launch
Science Challenge

Wanted: Space Investigator

You receive an email with the subject line "Wanted: Space Investigators." The email reads:

"The National Aeronautics and Space Administration (NASA) has selected you to manage a team of space investigators to analyze and model objects in space. Your team will deliver a presentation to government officials who want to know more about our galaxy and solar system, and the technology used to explore space."

In 3 ... 2 ... 1 ... begin your mission, Space Investigator!

Lesson 1
Gravity and the Universe

Lesson 2
The Solar System

Start Thinking About It

What properties would you include in a scale model of the solar system?

STEM Module Project

Planning and Completing the Science Challenge How will you meet this goal? The concepts you will learn throughout this module will help you plan and complete the Science Challenge. Just follow the prompts at the end of each lesson!

SCIENCE PROBES

LESSON 1 LAUNCH

Gravity in Space?

Olivia asked her friends what they know about gravity in space. This is what they said:

Justin: Gravity is a force that does not exist in space because objects float in space.

Taylor: I think gravity is a force that only exists on the surfaces of planets.

Juanita: Gravity is a force that exists between all objects that have mass—even in space!

Which friend do you agree with most? Explain why you agree.

You will revisit your response to the Science Probe at the end of the lesson.

LESSON 1

Gravity and the Universe

ENCOUNTER THE PHENOMENON | What would space be like without gravity?

GO ONLINE

Watch the video *Forming Galaxies* to see this phenomenon in action.

As you watch the video, refer back to the list of components in the universe that you compiled earlier. Add any new objects to the list as needed. After the video is over, discuss with a partner what you think would happen to these components if gravity did not exist.

EXPLAIN THE PHENOMENON

Are you starting to get some ideas about how gravity impacts components in solar and galaxy systems? Use your observations about the phenomenon to make a claim about the role of gravity in the universe.

CLAIM

Gravity plays a vital role in the universe because...

COLLECT EVIDENCE as you work through the lesson. Then return to these pages to record your evidence.

EVIDENCE

A. What evidence have you discovered to explain the factors that affect gravitational force?

B. What evidence have you discovered to explain how the solar system formed?

MORE EVIDENCE

C. What evidence have you discovered to explain how gravity impacts the formation and structure of galaxies?

When you are finished with the lesson, review your evidence. If necessary, based on the evidence, revise your claim.

REVISED CLAIM

Gravity plays a vital role in the universe because...

Finally, explain your reasoning for how and why your evidence supports your claim.

REASONING

The evidence I collected supports my claim because...

What is gravity?

Have you ever dropped your books while walking to your classroom? Or bounced a basketball in gym class? What keeps you, your books, a basketball, and everything else on Earth... on Earth?

INVESTIGATION

What Goes Up Must Come Down

Study the figure below. In models of gravitational force, strength is represented by the thickness of the arrows. The thicker the arrows, the greater the force.

The two objects in row A are the same distance apart as the two objects in row B. One of the objects in row B has more mass, creating a stronger gravitational force between the two objects in row B.

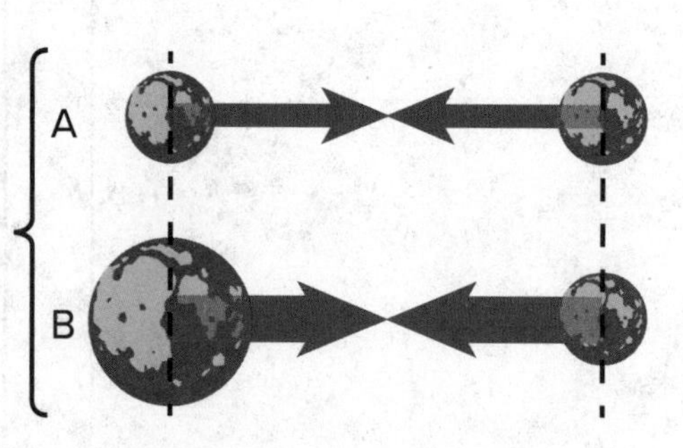

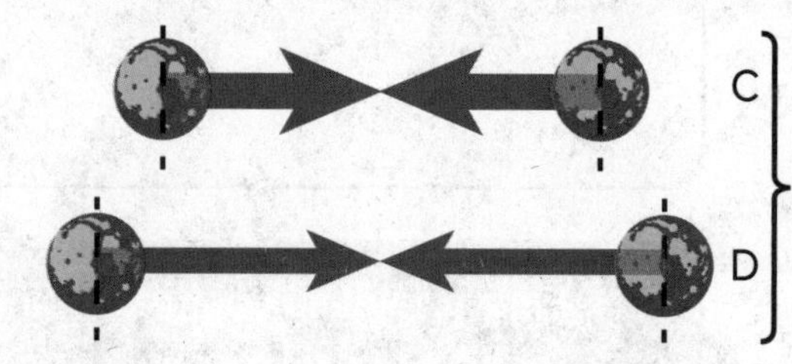

All four objects have the same mass. The two objects in row C are closer to each other than the two objects in row D and, therefore, have a stronger gravitational force between them.

1. What factors affect the strength of the gravitational force between objects?

2. Use the model above to construct an argument on why Earth exerts a greater gravitational force on you than other objects do.

Want more information?
Go online to read more about the role of gravity in the universe.

FOLDABLES®
Go to the Foldables® library to make a Foldable® that will help you take notes while reading this lesson.

The Force of Gravity Objects fall to the ground because Earth exerts an attractive force on them. **Gravity** is an attractive force that exists between all objects that have mass. The force of gravity between two objects depends on the objects' masses and the distance between them. The more mass either object has, or the closer together they are, the stronger the gravitational force.

Earth's gravity holds us on Earth's surface. Since Earth has more mass than any object near you, it exerts a greater gravitational force on you than other objects do. You don't notice the gravitational force between less massive objects.

THREE-DIMENSIONAL THINKING

1. Using what you have learned, create a diagram **modeling** why the Moon orbits Earth and not the Sun. Use force arrows to show gravitational relationships within the **system.**

2. Using your model as evidence, **argue** how gravitational force is dependent upon the masses of interacting objects and the distance between them. Construct your argument in your Science Notebook.

COLLECT EVIDENCE

What factors affect gravitational force? Record your evidence (A) in the chart at the beginning of the lesson.

What is gravity's role in the formation of stars?

Just as the force of gravity pulls a basketball to the floor, the force also pulls dust and gas inward in space. What impact does this have on objects in the universe, such as galaxies and stars? Let's find out.

PHYSICAL Connection When gas and dust clump together and form nebulae, the particles bump into each other and move faster. Gravity makes these nebulae hotter—as the particles are pulled closer together as a result of gravity, they start to bump into each and cause them to move faster, which causes the temperature to increase. The hotter something is, the more quickly its atoms move. As atoms move, they collide. If a gas is hot enough and its atoms move quickly enough, the nuclei of some of the atoms combine. Nuclear fusion is a process that occurs when the nuclei of several atoms combine into one larger nucleus. Once nuclear fusion has occurred, a star is formed.

What is the role of gravity in the formation of the solar system?

When the solar system formed, it was an enormous ball of gas and dust spinning slowly in space. As gravity pulled it closer together, the solar system spun faster. What happened to the shape of the solar system as it spun faster?

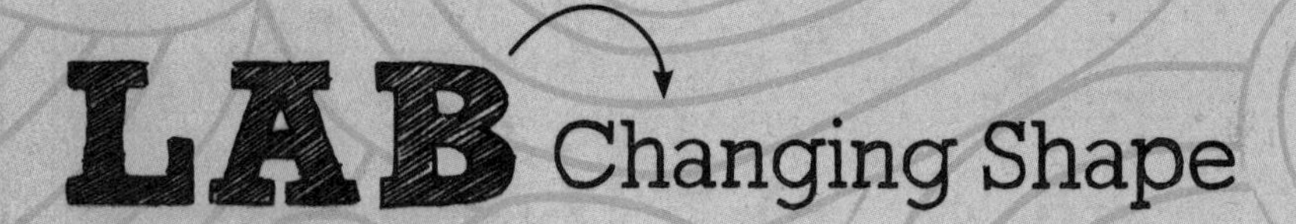

Safety 

Materials

salt dough
small bucket
stopwatch
string

Procedure

1. Read and complete a lab safety form.
2. Make a round ball about the size of your fist from a piece of salt dough.
3. Place the dough in a small bucket. Attach 1 m of sturdy string to the bucket's handle. **CAUTION:** *Be sure the string is securely attached.*

4. **CAUTION:** *Stand away from all furniture and people.* Whirl the bucket around your head for 1 min.
5. Lower the bucket. Observe the salt dough and record your observations in the Data and Observations section.

Data and Observations

Analyze and Conclude

6. What happened to the salt dough? What other objects change shape as they spin?

7. How do you think gravity influenced the shape of the early solar system?

Solar System Formation and Structure The solar system formed from a rotating cloud of gas, ice, and dust called a nebula. Gravity caused the cloud to collapse inward. Because the cloud was still rotating after the collapse, it flattened into a disk. Gravity pulled gas and dust toward the center of the disk, forming the Sun. Next, the planets began to form as gravity pulled these small particles together. As they collided, they stuck to each other and formed larger, unevenly shaped objects.

These larger objects had more mass and attracted more particles. Eventually enough matter collected and formed Earth and the seven other planets.

The leftover larger particles formed into moons that orbit the planets. Asteroids orbit the Sun in the asteroid belt, while meteoroids, dwarf planets, and comets orbit the Sun in their own paths through the solar system.

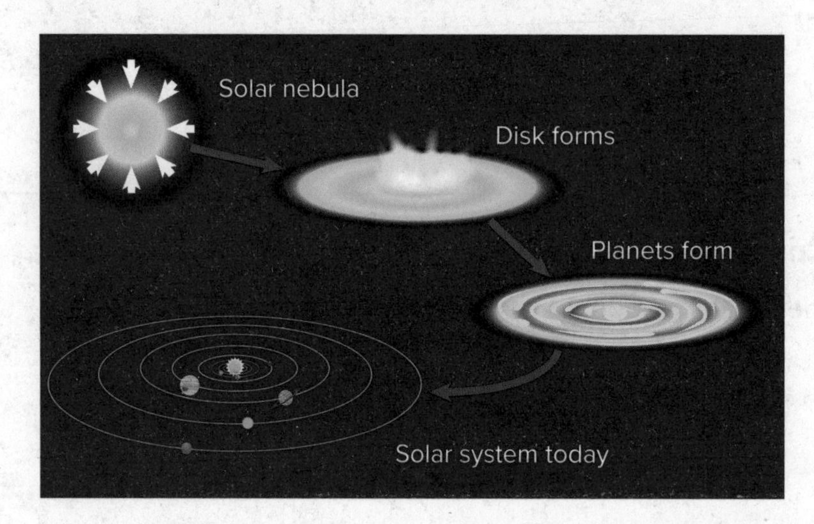

COLLECT EVIDENCE

How did the solar system form? Record your evidence (B) in the chart at the beginning of the lesson.

THREE-DIMENSIONAL THINKING

Using what you know about how the solar system formed, **construct an explanation** of how other solar systems in other galaxies form.

AMERICAN MUSEUM OF NATURAL HISTORY

CAREERS in SCIENCE

History from Space

Meteorites give a peek back in time.

More than 4.5 billion years ago, the planets of our solar system did not exist. Instead, a large disk of gas and dust, called the solar nebula, swirled around a young Sun. How did the planets and other objects in our solar system form from the solar nebula?

▲ **Denton Ebel holds a meteorite that was ejected from the asteroid 4 Vesta by a giant impact.**

Denton Ebel, a cosmochemist at the American Museum of Natural History in New York City, is investigating this question. Ebel wants to know the processes by which the first solid objects formed and how they combined to form larger bodies in our solar system. He looks for clues in certain kinds of meteorites called chondrites. Chondrites contain chondrules, clumps of dust grains that melted and solidified forming small glassy beads containing crystals. Chondrites also contain calcium-aluminum inclusions (called CAIs), the oldest rocks. Chondrules and CAIs are some of the oldest material in our solar system. Combined with a fine-grained mineral matrix, they make chondrites whose compositions are similar to that of the Sun but without the gases.

Ebel uses scientific instruments, such as a CAT-scan imager and an electron microprobe, to study chondrites. The instruments help him identify the abundance and composition of minerals in the chondrules. Using this data Ebel develops computer models based on the laws of physics and chemistry that simulate the conditions under which the chondrules, CAIs, and chondrites formed.

Through his research and collaboration with astrophysicists, Ebel can begin to understand how the solid precursors to planets formed, where in the solar nebula they formed, and why the rocky inner planets: Mercury, Venus, Mars and Earth, are different in composition from each other.

Nebular Theory

The solar system formed in stages.

- ***Gravity caused the solar nebula to collapse forming a rotating disk with the Sun at its center.***
- ***As the disk cooled, rocky material condensed in the hot inner regions. Ices condensed from gas in the cooler outer regions.***
- ***Tiny particles collided and accreted rapidly. Through repeated collisions they eventually formed the planets. Chondrites are leftovers of this process.***

▲ **These overlaid element maps of a single chondrule, taken with an electron microprobe, reveal magnesium-rich minerals in glass: red for magnesium, green for calcium, and blue for aluminum. Metal is black.**

It's Your Turn

TIME LINE Work in groups. Learn more about the history of Earth from its formation until life began to appear. Create a time line showing major events. Present your time line to the class.

How does gravity affect objects that orbit the Sun?

Have you ever swung a ball on the end of a string in a circle over your head? In some ways, the motion of a planet around the Sun is like the motion of that ball. As shown in the figure below, the Sun's gravitational force pulls each planet toward the Sun. This force is similar to the pull of the string that keeps the ball moving in a circle. The Sun's gravitational force pulls on each planet and keeps it moving along a curved path around the Sun.

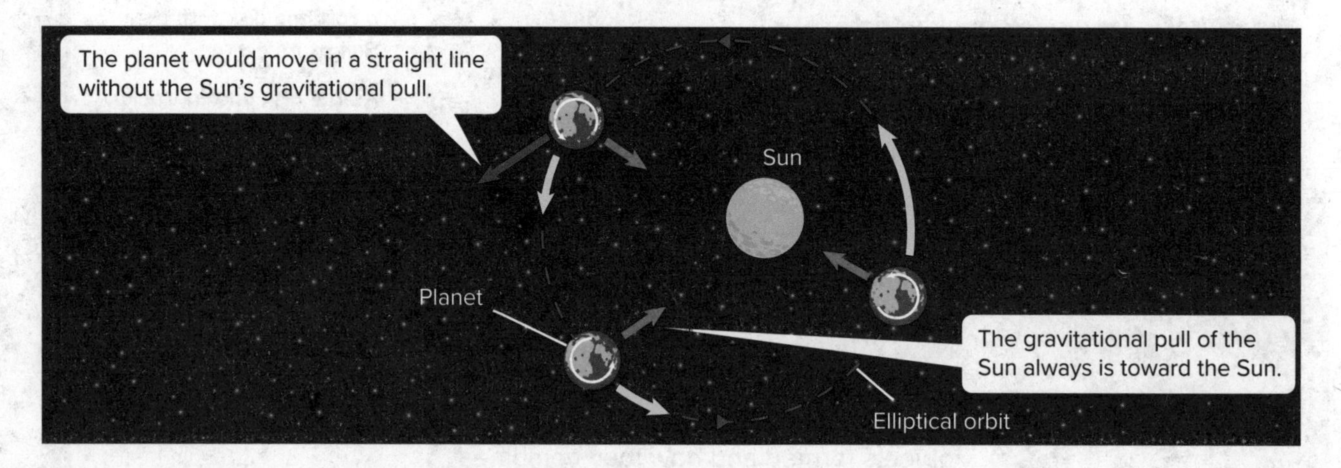

Unlike the ball on the string, planets do not move in circles. Instead, they move in an elliptical shape. Let's investigate those orbits.

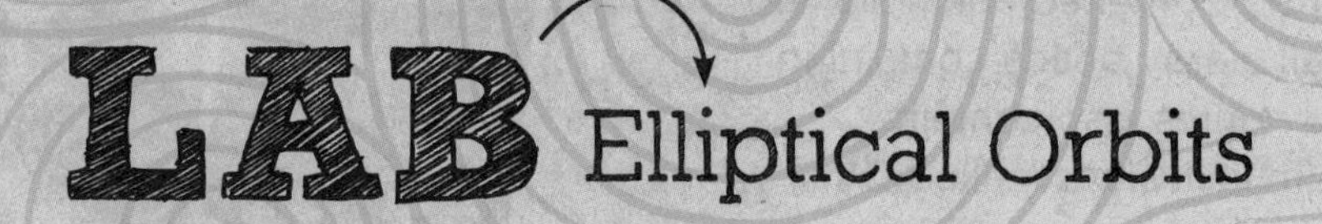

LAB Elliptical Orbits

Safety 

Materials

paper	string
cardboard	metric ruler
push pins (2)	scissors

Procedure

1. Read and complete a lab safety form.
2. Place a sheet of paper on a piece of cardboard. Insert two push pins 8 cm apart in the center of the paper.
3. Use scissors to cut a 24-cm piece of string. Tie the ends of the string together.

4. Place the loop of string around the pins. Use a pencil to draw an ellipse.
5. Measure the maximum width and length of the ellipse. Record your data in the Data and Observations section.
6. Move one of the push pins so that the pins are 5 cm apart. Repeat steps 4 and 5. Record your observations.

Data and Observations

Analyze and Conclude

7. Compare and contrast the two ellipses.

Planetary Orbits Earth was once thought to be the center of our solar system. In this geocentric model, the Sun, the Moon, and the planets revolved in circular orbits around a stationary Earth. In the early 1500s, Nicholas Copernicus proposed that Earth and other planets revolve in circular orbits around a stationary Sun, a heliocentric model.

HISTORY Connection In the 1600s, Johannes Kepler discovered that planets' orbits are ellipses, not circles. An ellipse contains two fixed points, called foci (singular, focus). Foci are equal distance from the ellipse's center and determine its shape. The Sun is at one focus. As a planet revolves, the distance between the planet and the Sun changes. Kepler also discovered that a planet's speed increases as it gets nearer to the Sun.

Why don't all celestial objects in the solar system orbit the Sun? Remember the gravitational force between objects depends on mass and distance. So, the Moon orbits Earth because the gravitational attraction between the Moon and Earth is stronger than the attraction between the Moon and the Sun.

What are galaxies?

Most people live in towns or cities where houses are close together. Not many houses are found in the wilderness. Similarly, most stars exist in galaxies. **Galaxies** are huge collections of gas, dust, and stars held together by gravity. The universe contains hundreds of billions of galaxies, and each galaxy can contain hundreds of billions of stars. And many of those stars have planets and other celestial objects orbiting them.

Groups of Galaxies Galaxies are not distributed evenly in the universe. Gravity holds them together in groups called clusters. Some clusters of galaxies are enormous. The Virgo Cluster contains about 2,000 galaxies. Most clusters exist in even larger structures called superclusters.

Virgo Cluster

The Milky Way One of those galaxies is our galaxy—the Milky Way. Our solar system and many other solar systems are in the Milky Way, a galaxy that contains almost 200 billion stars. These stars and solar systems orbit around the center of the Milky Way. The Milky Way is a member of the Local Group, a cluster of about 30 galaxies.

How are galaxies classified?

But are all galaxies just like the Milky Way? No, each galaxy is different from every other galaxy—no two galaxies are alike. Let's find out how they are different.

INVESTIGATION

Classification of Galaxies

GO ONLINE Watch the *Types of Galaxies* video to learn more about galaxies in the universe.

While watching the video, take notes in your Science Notebook of what you see.

How would you classify the galaxies you observed in the video?

Types of Galaxies In the video you observed three different types of galaxies—elliptical, spiral, and irregular. They are classified based on their shape.

Types of Galaxies

Spiral Galaxies

The stars, gas, and dust in a spiral galaxy exist in the spiral arms that begin at the central disk. Some spiral arms are long and symmetric; others are short and stubby. Spiral galaxies are thicker near the center, a region called the central bulge. A spherical halo of globular clusters and older, redder stars surrounds the disk.

Elliptical Galaxies

Unlike spiral galaxies, elliptical galaxies do not have spiral arms. Elliptical galaxies have a higher percentage of old, red stars. They contain little or no gas. Scientists suspect that many elliptical galaxies form by the gravitational merging of two or more spiral galaxies.

Irregular Galaxies

Irregular galaxies are oddly shaped. Many formed from the gravitational pull of neighboring galaxies. Irregular galaxies contain young stars and have areas of intense star formation.

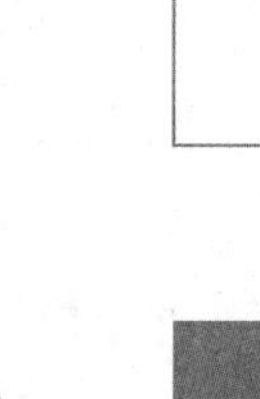

COLLECT EVIDENCE

How does gravity impact the formation and structure of galaxies? Record your evidence (C) in the chart at the beginning of the lesson.

LESSON 1

Review

Summarize It!

1. **Organize** Build a graphic organizer that shows how a solar system is formed. Make sure to include celestial objects such as stars, planets, moons, asteroids, meteoroids, and comets.

Three-Dimensional Thinking

It was once thought that Earth was the center of the universe. Eventually, it was proven that the planets orbit around the Sun. The illustration shows the path of Earth's orbit around the Sun.

2. Describe the path of Earth if the Sun's gravity were to suddenly stop.

A Earth would continue to move within its orbit.

B Earth would move in a straight line towards the Sun.

C Earth would move in a straight line instead of a curved line.

D Earth would stop moving and become suspended in one spot.

Halley's Comet orbits the Sun and can be seen from Earth about every 76 years. However, before the work of Sir Isaac Newton and Edmond Halley, comets were thought to pass in a straight line through the solar system. In 1705, Edmond Halley used Newton's laws to determine the gravitational effects of Jupiter and Saturn on a comet that he observed in 1682. Using this information and historical records, he determined that comets seen in 1531 and 1607 were the same comet. Halley correctly calculated the orbit of the comet and predicted its return in 1758.

3. Newton's laws state that all objects exert gravitational force and that objects with more mass exert more force. Which system of objects has the greatest effect on the orbit of Halley's comet?

A Earth, the Sun, and the Moon

B the Sun, Jupiter, and Saturn

C asteroids, meteoroids, and dwarf planets

D Earth, the Sun, and Saturn

Real-World Connection

4. Infer Scientists are actively searching for planets orbiting other stars. They can use gravity to find these planets, called exoplanets. Hypothesize how gravity helps locate distant planets orbiting distant stars.

5. Infer Earth has one orbiting moon. Jupiter has 79 confirmed moons. Infer why Earth has fewer moons than Jupiter.

Still have questions?
Go online to check your understanding about the role of gravity in the universe.

Do you still agree with the person you chose at the beginning of the lesson? Return to the Science Probe at the beginning of the lesson. Explain why you agree or disagree with that person now.

EXPLAIN THE PHENOMENON

Revisit your claim about how gravity impacts components in space. Review the evidence you collected. Explain how your evidence supports your claim.

START PLANNING

STEM Module Project
Science Challenge

Now that you learned about gravity and its importance in the universe, go to your Module Project to start analyzing the objects in the solar system and the universe and start planning your model.

LESSON 2 LAUNCH

Objects in Our Solar System

A system consists of parts that make up a whole. What are the different parts that make up our solar system? Put an X next to each of the objects you think are part of our solar system.

_____ planets	_____ the Sun	_____ nearby stars other than the Sun
_____ distant stars	_____ constellations	_____ asteroids
_____ comets	_____ moons	_____ human-made satellites
_____ galaxies	_____ black holes	_____ universe

Explain your thinking. Describe what determines whether an object is part of our solar system.

____

____

____

____

____

You will revisit your response to the Science Probe at the end of the lesson.

LESSON 2

The Solar System

ENCOUNTER THE PHENOMENON | How do objects in the solar system compare?

GO ONLINE

Watch the animation *In Our Neighborhood* to to see this phenomenon in action.

The animation shows how scientists organize objects in our solar system using distances. Think of other characteristics that you could use to organize the objects in our solar system. Create a graphic organizer showing your ideas about how objects in the solar system might be compared in the space below.

EXPLAIN THE PHENOMENON

You examined an animation about the solar system. How do astronomers learn about the objects within the solar system? How do planets, moons, asteroids, and comets compare to each other? Make a claim on how objects in the solar system can be analyzed and interpreted.

CLAIM

The objects of the solar system can be analyzed and interpreted by...

COLLECT EVIDENCE as you work through the lesson. Then return to these pages to record your evidence.

EVIDENCE

A. What evidence have you discovered for how exploration has helped the advancement of technology?

B. What evidence have you discovered for how the planets in the solar system compare to each other?

MORE EVIDENCE

C. What evidence have you discovered for how moons, asteroids, comets, meteors, and dwarf planets compare to each other?

When you are finished with the lesson, review your evidence. If necessary, based on the evidence, revise your claim.

REVISED CLAIM

The objects of the solar system can be analyzed and interpreted by...

Finally, explain your reasoning for how and why your evidence supports your claim.

REASONING

The evidence I collected supports my claim because...

What objects make up the solar system?

Recall that gas, dust, and stars make up galaxies. Stars are made up of gas that is undergoing fusion at the core. What's in a solar system? A solar system contains at least one star and all the objects that orbit around that star. This includes planets, moons, comets, asteroids, and meteoroids.

Want more information?
Go online to read more about the objects in the solar system and universe.

FOLDABLES®
Go to the Foldables® library to make a Foldable® that will help you take notes while reading this lesson.

How do astronomers observe the solar system?

Ancient observers looking at the night sky saw many stars but only five planets—Mercury, Venus, Mars, Jupiter, and Saturn. The invention of the telescope in the 1600s led to the discovery of additional planets and many other space objects.

Telescopes enable astronomers to observe many more stars than they could with their eyes alone. Telescopes are designed to collect a certain type of electromagnetic wave. Some telescopes detect visible light, and others detect radio waves and microwaves. Astronomers use many kinds of telescopes to study the energy emitted by stars and other objects in space. Let's look at the differences in telescope technology.

INVESTIGATION

Compare the View

Examine the images below.

Jupiter

Saturn

1. These images were taken from a telescope you might use in your backyard. Based on these photos, describe the differences between the two planets.

Now examine the images below.

Jupiter

Saturn

2. These images were taken from a space-based telescope. Based on these photos, describe the differences between the two planets.

Technology Advances In addition to telescopes, space agencies like NASA have sent probes and other exploratory spacecraft into space. For example, the *New Horizons* mission explored the Pluto system within our solar system and will explore objects on the edge of the solar system in the Kuiper belt. Other spacecraft, like the *International Space Station* (*ISS*), house rotating shifts of astronauts for research purposes. The experiments conducted on board the *ISS* will aid future crewed space flights.

The space program requires the development of materials that can withstand the extreme temperatures and pressures of space. Much of this new technology has also been adapted for use in materials on Earth. Some examples of everyday materials that were developed as a result of the space program are satellite television, medical imaging, and home insulation.

COLLECT EVIDENCE

How has space exploration helped the advancement of technology? Record your evidence (A) in the chart at the beginning of the lesson.

How do scientists analyze data about the solar system?

Knowing how scientists obtain data and information on objects, they can then analyze the information and determine the relationships and patterns between characteristics. Let's see how scientists analyze data from the planets in the solar system.

INVESTIGATION

Graphing Characteristics

Scientists collect and analyze data and draw conclusions based on data. Scientists know that some properties of the planets are related. Graphing data makes the relationships easy to identify. Graphing allows different types of data be to seen in relation to one another.

Scientists use different units of measurements for distance within the solar system. One of those units is the astronomical unit (AU). One AU is equal to the average distance between the Sun and Earth, about 149,600,000 km.

Planet	Average Distance from the Sun (AU)	Orbital Period (yr)	Radius of Planet (km)
Mercury	0.39	0.24	2,440
Venus	0.72	0.62	6,051
Earth	1.00	1.0	6,378
Mars	1.52	1.9	3,397
Jupiter	5.20	11.9	71,492
Saturn	9.58	29.4	60,268
Uranus	19.2	84.0	25,559
Neptune	30.1	164.0	24,764

1. You will plot two graphs that explore the relationships in data. The first graph compares a planet's distance from the Sun and its orbital period. The second graph compares a planet's distance from the Sun and its radius. Make a prediction about how these two sets of data are related, if at all. The data are shown in the table.

2. Use the data in the table to plot a line graph showing orbital period versus average distance from the Sun. On the *x*-axis, plot the planet's distance from the Sun. On the *y*-axis, plot the planet's orbital period. Make sure the range of each axis is suitable for the data to be plotted and clearly label each planet's data point.

Orbital Period v. Distance from the Sun

3. Examine the *Orbital Period v. Distance from the Sun* graph. Describe any relationship you observe between a planet's distance from the Sun and its orbital period.

4. Use the data in the table to plot a line graph showing planet radius versus average distance from the Sun. On the *x*-axis, plot the planet's distance from the Sun. On the *y*-axis, plot the planet's radius. Make sure the range of each axis is suitable for the data to be plotted and clearly label each planet's data point.

Planet Radius v. Distance from the Sun

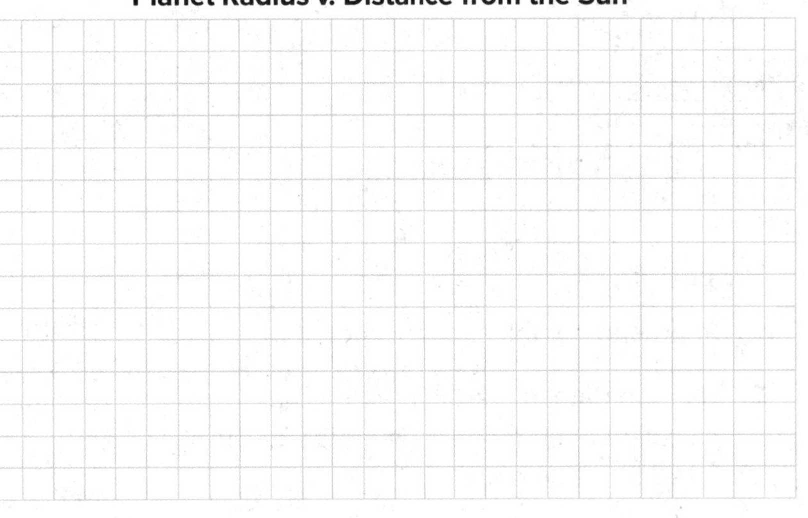

5. Examine the *Planet Radius v. Distance from the Sun* graph. Describe any relationship you observe between a planet's distance from the Sun and its radius. How does this compare to your prediction?

Orbits and Distance from the Sun You've learned that graphing data will sometimes show the relationships between the measured characteristics. The planets' orbital periods are directly related to the average distance from the Sun. You also discovered by graphing data that there is no relationship between the distance from the Sun and the radius of the planet.

How can you model the solar system?

A scale model is a physical representation of something that is much smaller or much larger. Reduced-sized scale models are used to represent and study very large things, such as the solar system. The scale used must reduce the actual size to a size reasonable for the model. Let's investigate how to use a scale model.

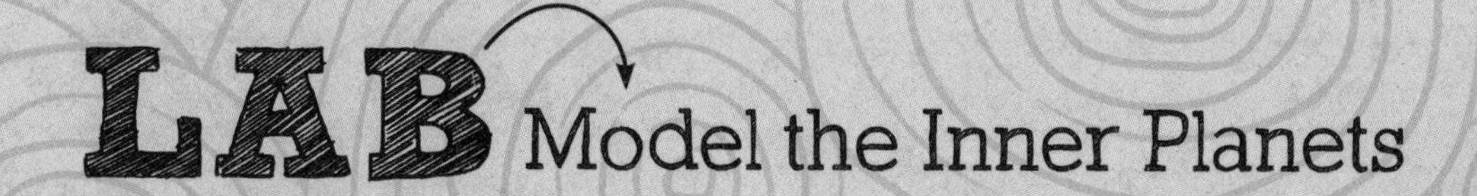

LAB Model the Inner Planets

Safety 

Materials

modeling clay

metric ruler

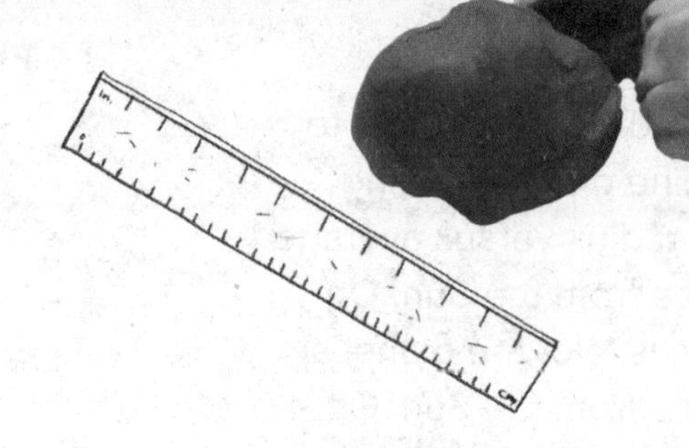

Procedure

1. Read and complete a lab safety form.
2. Use the data in the table for Earth to calculate each model's diameter for the other three planets.
3. Use modeling clay to make a ball that represents the diameter of each planet. Check the diameter with a metric ruler.

Planet	Actual Diameter (km)	Model Diameter (cm)
Mercury	4,879	
Venus	12,103	
Earth	12,756	8.0
Mars	6,792	

Analyze and Conclude

4. **MATH Connection** Explain how you converted actual diameters (km) to model diameters (cm).

5. How do the inner planets' diameters compare?

Now that you have modeled the diameter of the inner planets, let's model the distances in the solar system.

LAB Scale Down

Objects in the solar system are so far apart that astronomers use a larger distance unit. An astronomical unit (AU) is the average distance from Earth to the Sun—about 150 million km. Let's put this scaled measurement into action.

Safety

Materials

register tape   meterstick   masking tape

Procedure

1. Read and complete a lab safety form.
2. First, decide the scale of your solar system. Use the data given in the table on the next page to figure out how far apart the Sun and Neptune would be if a scale of 1 meter = 1 AU is used. Would a solar system based on that scale fit in the space you have available?
3. With your group, determine the scale that results in a model that fits the available space. Larger models are usually more accurate, so choose a scale that produces the largest model that fits in the available space.

Procedure, continued

4. **MATH Connection** Once you have decided on a scale, record the unit you have chosen in the third column of the table. Then fill in the scaled distance for each planet.

Planet	Distance from the Sun (AU)	Distance from the Sun (________________)
Mercury	0.39	
Venus	0.72	
Earth	1.00	
Mars	1.52	
Jupiter	5.20	
Saturn	9.54	
Uranus	19.18	
Neptune	30.06	

5. On register tape, mark the positions of the objects in the solar system based on your chosen scale. Use a length of register tape that is slightly longer than the scaled distance between the Sun and Neptune.

6. Tape the ends of the register tape to a table or the floor. Mark a dot at one end of the paper to represent the Sun. Measure along the tape from the center of the dot to the location of Mercury. Mark a dot at this position and label it *Mercury*. Repeat this process for the remaining planets.

Analyze and Conclude

7. There are many objects in the solar system. These objects have different sizes, structures, and orbits. Examine your scale model of the solar system. How accurate is the model? How could the model be changed to be more accurate?

8. Pluto is a dwarf planet located beyond Neptune. Based on the pattern of distance data for the planets shown in the table, approximately how far from the Sun would you expect to find Pluto? Explain your reasoning.

9. Compare your model with other groups in your class by taping them all side-by-side. Discuss any major differences in your models. Discuss the difficulties in making the scale models much smaller.

SEP DCI CCC

THREE-DIMENSIONAL THINKING

How can you build a **scale model** of the solar system that accurately shows both planetary diameters and distances and the locations of the asteroid belt and dwarf planets? Describe how you would create this model.

INVESTIGATION

Digging Deeper

You have created a model of the planetary distances in our solar system. As a class, you will determine the scale diameter of each of the planets and then research and create a model of their inner layers and the composition of those layers.

Work together and choose a planet to model. After you've chosen a planet, research that planet and its structure and composition. Take notes about your planet below.

1. What is your planet composed of?

2. How is your planet structured?

3. What surface features are present on your planet?

Using materials of your choice, begin constructing your model. You may also create a model on a computer program that can create a 3-D model.

Once you have created your model, share your model with the class. Make sure you are able to answer questions about the surface features and composition of your planet.

COLLECT EVIDENCE

How do the planets in the solar system compare to each other? Record your evidence (B) in the chart at the beginning of the lesson.

Read a Scientific Text

Astronomers learn more about the terrestrial planets using space probes.

CLOSE READING

Inspect

Read the passage *The Incredible Shrinking Mercury is Active After All.*

Find Evidence

Reread the fourth paragraph. Underline the surface features that are present on Mercury.

Make Connections

Communicate With your partner, research surface features on Venus and Mars and compare those with the surface features on Earth and Mercury.

PRIMARY SOURCE

The Incredible Shrinking Mercury is Active After All

It's small, it's hot, and it's shrinking. New NASA-funded research suggests that Mercury is contracting even today, joining Earth as a tectonically active planet.

Images obtained by NASA's MErcury Surface, Space ENvironment, GEochemistry, and Ranging (MESSENGER) spacecraft reveal previously undetected small fault scarps—cliff-like landforms that resemble stair steps. These scarps are small enough that scientists believe they must be geologically young, which means Mercury is still contracting and that Earth is not the only tectonically active planet in our solar system, as previously thought. ...

"The young age of the small scarps means that Mercury joins Earth as a tectonically active planet, with new faults likely forming today as Mercury's interior continues to cool and the planet contracts," said lead author Tom Watters, Smithsonian senior scientist at the National Air and Space Museum in Washington, D.C.

Large fault scarps on Mercury were first discovered in the flybys of Mariner 10 in the mid-1970s and confirmed by MESSENGER, which found the planet closest to the sun was shrinking. The large scarps were formed as Mercury's interior cooled, causing the planet to contract and the crust to break and thrust upward along faults making cliffs up to hundreds of miles long and some more than a mile (over one-and-a-half kilometers) high.

In the last 18 months of the MESSENGER mission, the spacecraft's altitude was lowered, which allowed the surface of Mercury to be seen at much higher resolution. These low-altitude images revealed small fault scarps that are orders of magnitude smaller than the larger scarps. The small scarps had to be very young, investigators say, to survive the steady bombardment of meteoroids and comets. They are comparable in scale to small, young lunar scarps that are evidence Earth's moon is also shrinking.

This active faulting is consistent with the recent finding that Mercury's global magnetic field has existed for billions of years and with the slow cooling of Mercury's still hot outer core. It's likely that the smallest of the terrestrial planets also experiences Mercury-quakes—something that may one day be confirmed by seismometers. ...

Source: National Aeronautics and Space Administration

What are the other objects in the solar system?

Different types of objects orbit the Sun. These objects include dwarf planets, asteroids, meteors, and comets. Unlike the Sun, these objects don't emit light but only reflect the Sun's light. What are the features, compositions, and locations of these objects? Let's investigate!

INVESTIGATION

Moons of the Outer Planets

Jupiter has 79 confirmed moons. Saturn has 53 named moons and 8 provisional moons. Uranus has 27 moons while Neptune only has 13 named moons with one provisional moon. Why do the outer planets have so many moons compared to the inner planets?

The moons of the outer planets range in diameter from 2 km to 5,268 km. The largest moon in the solar system is Jupiter's Ganymede, which is larger than the planet Mercury. How can a moon be larger than a planet? Remember, to be classified as a moon, an object has to orbit a planet, and not the Sun.

Moons of the Outer Planets			
Planet	**Number of Moons**	**Largest Moons**	**Moon Sizes**
Jupiter	79 confirmed	Ganymede, Callisto, Io, Europa	These four moons are planet-sized. Ganymede is the solar system's largest moon, with a diameter of 5,268 km.
Saturn	at least 53	Titan, Rhea, Iapetus, Dione	Titan is planet-sized at 5,150 km in diameter.
Uranus	at least 27	Titania, Oberon, Umbriel, Ariel	Titania is 1,578 km in diameter, while tiny Cordelia is 26 km in diameter.
Neptune	at least 13	Triton, Proteus, Nereid, Larissa	Triton is 2,700 km in diameter.

1. Select the largest moon of each outer planet. Then, draw the moons in order from largest to smallest. Identify each moon and its planet. Create a scale so you can keep your drawings proportional. Be sure to show your scale.

2. With a partner, research each of the identified moons. Compare and contrast a chosen characteristic—surface features, composition, structure, distance from the planet—for each of the moons. Use your Science Notebook if you need more room.

Moons Many moons of the outer planets are small with irregular shapes and unusual orbits. Some scientists think that these are captured asteroids, meaning the gravity of the planet pulled the object into its orbit. Captured asteroids are natural satellites, but they did not form by accretion as regular spherical moons probably did.

Dwarf Planets Scientists classify some objects in the solar system as dwarf planets. A **dwarf planet** is a spherical object that orbits the Sun. It is not a moon of another planet and is in a region of the solar system where there are many objects orbiting near it. But, unlike a planet, a dwarf planet does not have more mass than objects in nearby orbits. Dwarf planets include Ceres (SIHR eez), Eris (IHR is), Pluto, and Makemake (MAH kay MAH kay). Dwarf planets are made of rock and ice and are much smaller than Earth.

Other Solar System Objects	
Asteroids **Asteroids** are small, rocky objects that never clumped together like the rocks and ice that formed the inner planets. Some astronomers suggest that Jupiter's strong gravity might have caused the chunks to collide so violently that they broke apart instead of sticking together. This means that asteroids are objects left over from the formation of the solar system. Most asteroids are found in the asteroid belt between Mars and Jupiter.	
Comets **Comets** are mixtures of rock, ice, and dust. The particles in a comet are loosely held together by the gravitational attractions among the particles. Comets orbit the Sun in long elliptical orbits.	
Meteoroids, Meteors, Meteorites **Meteoroids** are small, rocky particles that move through space. Most meteoroids are only about as big as a grain of sand. When meteoroids pass through Earth's atmosphere, they burn up because of friction and are seen as streaks of light in the sky. This is called a **meteor.** If a meteor doesn't completely burn up, it can impact the planet's surface. At that point, it is called a **meteorite.**	

COLLECT EVIDENCE

How do the moons, asteroids, comets, meteors, and dwarf planets compare to each other? Record your evidence (C) in the chart at the beginning of the lesson.

GO ONLINE for additional opportunities to explore!

Investigate more about dwarf planets and other objects in the solar system by completing one or both of these activities.

☐ **Read** the **Scientific Text** *Dwarf Planets and Other Objects—Comets.*

OR

☐ **Complete** the **Lab** *How might asteroids and moons form?*

AMERICAN MUSEUM of NATURAL HISTORY

Pluto

CAREERS in SCIENCE

What in the world is it?

Since Pluto's discovery in 1930, students have learned that the solar system has nine planets. But in 2006, the number of planets was changed to eight. What happened?

Astrophysicists at the American Museum of Natural History in New York City were among those that first questioned Pluto's classification as a planet. Pluto orbits the Sun at the outer edge of the solar system in a region called the Kuiper Belt. It shares this region with thousands of other objects. One of the largest objects, known as Eris, is almost the same size as Pluto. Shouldn't Eris be classified as a planet too?

This question spurred the International Astronomical Union (IAU) to set up a committee to define what constitutes a planet. In 2006, the IAU came up with three rules for being considered a planet in our solar system. First, the object has to be massive enough for gravity to shape it into a sphere. Secondly, it has to be in orbit around the Sun. Finally, the object must have a clear orbit devoid of other objects. In other words, it has to have its own "zone." Because Pluto has other objects in its zone, it was demoted from its status as a planet and reclassified as a dwarf planet. And while Pluto may have lost its rank as the smallest planet, it still is "King of the Kuiper Belt."

Pluto TIME LINE

1930
Astronomer Clyde Tombaugh discovers a ninth planet, Pluto.

1992
The first Kuiper Belt object is discovered.

2005
Eris—an object the size of Pluto—is discovered in the Kuiper Belt.

January 2006
NASA launched the *New Horizons* spacecraft that will study Pluto, its moon, and other Kuiper Belt objects.

August 2006
Pluto is reclassified as a dwarf planet.

July 2015
New Horizons reached Pluto and began gathering data.

This illustration shows what Pluto might look like if you were standing on one of its moons.

It's Your Turn

MAKE A LIST What did scientists learn about Pluto when the *New Horizons* spacecraft visited the Pluto system in 2015? Make a list with three things they discovered.

LESSON 2

Review

Summarize It!

1. **Construct** a graphic organizer that compares and contrasts objects in the solar system.

Three-Dimensional Thinking

NASA has been sending exploration missions to Mars for almost 50 years. The length of time it has taken the different spacecraft to reach Mars ranges from 150 days to 360 days. In order to make it possible for humans to travel to Mars, the travel time must be reduced.

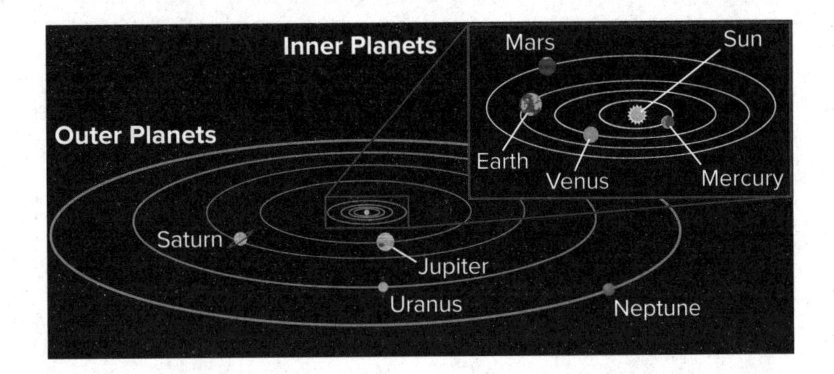

2. Assume the different spacecraft were using the same amount of fuel at the same rate. Why are there such long differences in travel time?

A The distance of Mars from the Sun changes.

B The distance from Earth to Mars changes.

C The position of Earth from the Sun changes.

D The shape of the planet's orbit changes.

The chart below shows the surface temperatures of the outer planets.

Planet	Surface Temperature
Neptune	–214°C
Uranus	–216°C
Jupiter	–148°C
Saturn	–178°C

3. What can you conclude about the outer planets based on the data table?

A If Neptune were closer to the Sun, it would be cooler.

B Saturn is closer to the Sun than any of the other outer planets.

C The coldest planet is Uranus.

D The surface temperature of Earth is between that of Neptune and Jupiter.

Real-World Connection

4. **Evaluate** Think about the ways in which scientists observe, study, and explore space. What technology, other than the examples given to you in this lesson, has been added to everyday life because of space exploration?

Still have questions?

Go online to check your understanding about objects in our solar system.

REVISIT

Do you still agree with the answers you chose at the beginning of the lesson? Return to the Science Probe at the beginning of the lesson. Explain why you agree or disagree with those answers now.

EXPLAIN THE PHENOMENON

Revisit your claim about how objects in our solar system can be analyzed and interpreted. Review the evidence you collected. Explain how your evidence supports your claim.

PLAN AND PRESENT

STEM Module Project
Science Challenge

Now that you learned about the objects in the solar system, go to your Module Project to continue analyzing the objects in the solar system and the universe, finish building your model, and give your presentation.

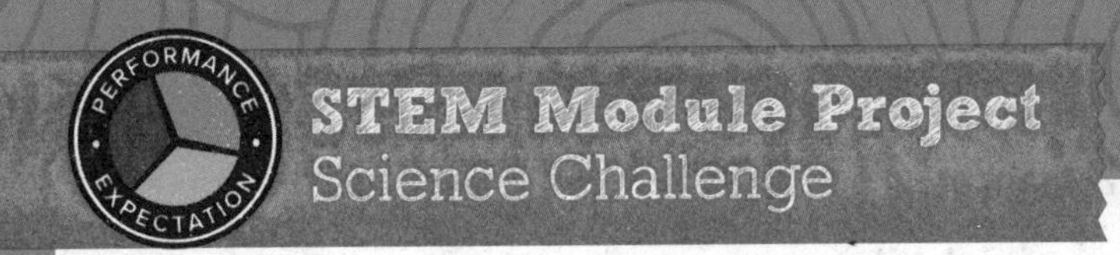

Wanted: Space Investigator

You receive an email with the subject line "Wanted: Space Investigators." The email reads:

"The National Aeronautics and Space Administration (NASA) has selected you to manage a team of space investigators to analyze and model objects in space. Your team will deliver a presentation to government officials who want to know more about our galaxy and solar system, and the technology used to explore space."

In 3 ... 2 ... 1 ... begin your mission, Space Investigator!

Planning After Lesson 1

How can you indicate the relative spatial scales of our solar system and galaxy in a model?

Gather the data you need for your model and organize it in your Science Notebook. Cite the sources of your data.

How will you incorporate the role of gravity in your model?

Planning After Lesson 2

Review the data you gathered in Lesson 2 about objects in our solar system. What additional data do you need for your model and analysis of solar system objects?

Gather your data and organize it in the space below. Be sure to cite the sources of your data. Use your Science Notebook if you need more space.

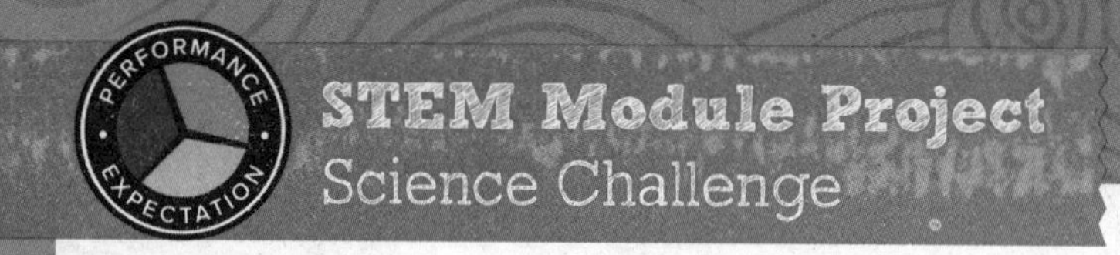

Develop Your Model

Review the planning you did after Lessons 1 and 2, then develop a model that describes the role of gravity in the solar system and Milky Way galaxy. You will also need to analyze and interpret your data in order to incorporate elements in your model that show the scale properties of objects in our solar system. Sketch your model below. List the materials you will use.

Evaluate Your Model

Model Elements	Descriptions
Components (What are the different parts of my model?)	
Relationships (How do the components of my model interact?)	
Connections (How does my model help me understand the phenomenon?)	

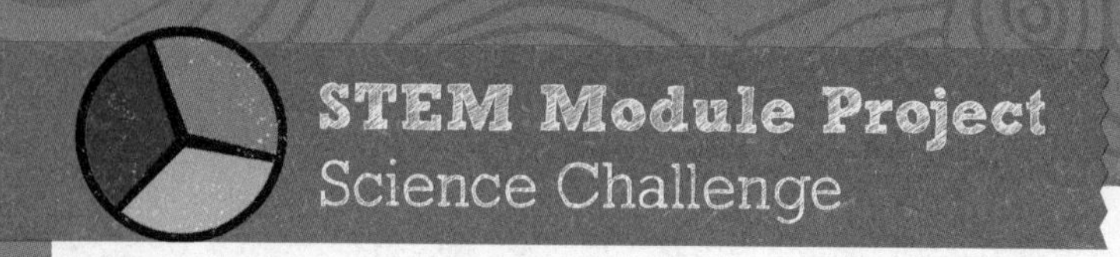

Present Your Model

After you have observed and evaluated your model, answer the questions below.

Based on your data, identify one advance in solar system science made possible by improved engineering. Conversely, identify one advance in engineering made possible by advances in solar system science.

Use your model to identify patterns in the characteristics of solar system objects. Describe one pattern below.

Use your model to identify patterns or draw conclusions about the scale of solar system objects. Write one pattern or conclusion below.

Congratulations! You've completed the Science Challenge requirements.

Module Wrap-Up

REVISIT THE PHENOMENON

Using the model and analysis of the objects in the solar system, explain the relationship between solar system objects and other objects in the universe.

OPEN INQUIRY

If you had to ask one question about what you studied, what would it be?

Plan and conduct an investigation to answer this question.

Glossary

Multilingual Glossary

GO ONLINE to find multilingual glossaries for science.
The glossaries include the following languages.

Arabic	Korean	Tagalog
Bengali	Mandarin Chinese	Urdu
French	Portuguese	Vietnamese
Haitian Creole	Russian	
Hmong	Spanish	

Cómo usar el glosario en español:

1. Busca el término en inglés que desees encontrar.
2. El término en español, junto con la definición, se encuentran en la columna de la derecha.

Pronunciation Key

Use the following key to help you sound out words in the glossary.

a	back (BAK)	**Ew**	food (FEWD)
ay	day (DAY)	**yoo**	pure (PYOOR)
ah	father (FAH thur)	**yew**	few (FYEW)
ow	flower (FLOW ur)	**uh**	comma (CAH muh)
ar	car (CAR)	**u (+ con)**	rub (RUB)
E	less (LES)	**sh**	shelf (SHELF)
ee	leaf (LEEF)	**ch**	nature (NAY chur)
ih	trip (TRIHP)	**g**	gift (GIHFT)
i (i + com + e)	idea (i DEE uh)	**J**	gem (JEM)
oh	go (GOH)	**ing**	sing (SING)
aw	soft (SAWFT)	**zh**	vision (VIH zhun)
or	orbit (OR buht)	**k**	cake (KAYK)
oy	coin (COYN)	**s**	seed, cent (SEED)
oo	foot (FOOT)	**z**	zone, raise (ZOHN)

English — **A** — **Español**

asteroid: a small, rocky object that orbits the Sun.

asteroide: objeto pequeño y rocoso que orbita el Sol.

C

comet: a small, rocky, icy object that orbits the Sun.

cometa: objeto pequeño, rocoso y helado que orbita el Sol.

D

dwarf planet: an object that orbits the Sun and is nearly spherical in shape, but shares its orbital path with other objects of similar size.

planeta enano: objeto de forma casi esférica que orbita el Sol y que comparte el recorrido de la órbita con otros objetos de tamaño similar.

E

equinox: when Earth's rotation axis is tilted neither toward nor away from the Sun.

equinoccio: cuando el eje de rotación de la Tierra se inclina sin acercarse ni alejarse del Sol.

G

galaxy: a huge collection of stars, gas, and dust.

galaxia: conjunto enorme de estrellas, gas, y polvo..

gravity: an attractive force that exists between all objects that have mass.

gravedad: fuerza de atracción que existe entre todos los objetos que tienen masa.

L

lunar eclipse: an occurrence during which the Moon moves into Earth's shadow.

eclipse lunar: ocurrencia durante la cual la Luna se mueve hacia la zona de sombra de la Tierra.

M

meteor: a meteoroid that has entered Earth's atmosphere and produces a streak of light.

meteorite: a meteoroid that strikes a planet or a moon.

meteoroid: a small, rocky particle that moves through space.

meteoro: meteorito que ha entrado a la atmósfera de la Tierra y produce un haz de luz.

meteorito: meteoroide que impacta un planeta o una luna.

meteoroide: partícula rocosa pequeña que se mueve por el espacio.

O

orbit: the path an object follows as it moves around another object.

órbita: trayectoria que un objeto sigue a medida que se mueve alrededor de otro objeto.

P

penumbra: the lighter part of a shadow where light is partially blocked.

phase: the lit part of the Moon or a planet that can be seen from Earth.

penumbra: parte más clara de una sombra donde la luz se bloquea parcialmente.

fase: parte iluminada de la Luna o de un planeta que se ve desde la Tierra.

R

revolution: the orbit of one object around another object.

rotation: the spin of an object around its axis.

rotation axis: the line on which an object rotates.

revolución: movimiento de un objeto alrededor de otro objeto.

rotación: movimiento giratorio de un objeto sobre su eje.

eje de rotación: línea sobre la cual un objeto rota.

S

solar eclipse: an occurrence during which the Moon's shadow appears on Earth's surface.

solstice: when Earth's rotation axis is tilted directly toward or away from the Sun.

eclipse solar: acontecimiento durante el cual la sombra de la Luna aparece sobre la superficie de la Tierra.

solsticio: cuando el eje de rotación de la Tierra se inclina acercándose o alejándose del Sol.

U

umbra: the central, darker part of a shadow where light is totally blocked.

umbra: parte central más oscura de una sombra donde la luz está completamente bloqueada.

W

waning phases: phases of the Moon during which less of the Moon's near side is lit each night.

waxing phases: phases of the Moon during which more of the Moon's near side is lit each night.

fases menguantes: fases de la Luna durante las cuales el lado cercano de la Luna está menos iluminado cada noche.

fases crecientes: fases de la Luna durante las cuales el lado cercano de la Luna está más iluminado cada noche.

Italic numbers = illustration/photo
Bold numbers = vocabulary term
lab = indicates entry is used in a lab
inv = indicates entry is used in an investigation
smp = indicates entry is used in a STEM Module Project
enc = indicates entry is used in an Encounter the Phenomenon
sc = indicates entry is used in a STEM Career

A

Antarctic Circle, 21
Aphelion, 20
Apollo program, 37
Apparent motion, 12, *17*
Arctic Circle, 21
Asteroids, *108*
Astronomical seasons, 19

C

Chondrites, 83
Climatologists, 19
Comets, 108
Conversion equations, 20
Copernicus, Nicholas, 85
Cosmochemists, 83

D

Data analysis, 98–99 *inv*
Distance
gravity and, 78, *79*
planetary orbits and, 98–99 *inv*, 100
Dwarf planets, 107, 109

E

Earth
curvature of, 14
motion of, 10
orbit of, 11
polar v. equatorial regions, 14
revolution of, 11
rotation of, 11, 12
tilt of, 14, 15
Earth's Motion Around the Sun, 5–24
Ebel, Denton, 83
Eclipses
lunar, *52*
lunar phases and, 52–53, 54–55 *inv*
models of, 4 *smp*
patterns and, 54, 56
solar, 48, 49–50 *lab*, *51*
total v. partial, 51–52, 53
Eclipses, 41–60
Electromagnetic waves, 96
Elliptical galaxies, *87*
Elliptical orbits, 84–85 *lab*, 85
Encounter the Phenomenon, 1, 7, 27, 43, 71, 75, 93
Energy, 14, 15
Equinox, 17, *18*, 19. *see also* **Solstice**
Explain the Phenomenon, 8–9, 24, 28–29, 40, 44–45, 69, 76–77, 90, 94–95, 112, 119
Eye safety, 57

F

Foci, 85

G

Galaxies
defined, **86**
types of, 86 *inv*, 87
Galaxy clusters, 86
Geocentric model, 85
Graphing, 98–99 *inv*
Gravitational force, *79*, 84
Gravitational pull, 11, 31
Gravity
defined, **78**
distance and, 79
formation of the solar system and, 80–81 *lab*, 82, 83
galaxies and, 86
mass and, 79
in space, 80
Gravity and the Universe, 73–90

H

Heliocentric model, 85

I

International Astronomical Union (IAU), 109
***International Space Station*,** 97
Investigation
Ahead of the Curve, 13–14
Classification of Galaxies, 86
Compare the View, 96–97
Digging Deeper, 104
Eclipses, 54–55
Foil Moon, 30
Graphing Characteristics, 98–99
Moons of the Outer Planets, 106–107
Night and Day, 10
Star Gazing, 12
The Motion of the Moon, 31–32
What Goes Up Must Come Down, 78
Irregular galaxies, *87*

K

Kepler, Johannes, 85
Kuiper belt, 97, 109

L

Lab
Beyond a Shadow of a Doubt, 46–47
Casting Shadows, 49–50
Changing Shape, 80–81
Elliptical Orbits, 84–85
Model the Inner Planets, 100–101
Moon Phases, 33–35
Scale Down, 101–103
Law of Universal Gravitation, 11, 78
Leap Year, 19
Lunar Crater Observation and Sensing Satellite (LCROSS), 37
Lunar eclipses
described, *52*
lunar phases and, 52–53, 54–55 *inv*
total v. partial, 53
Lunar Phases, 25–40
eclipses and, 52–53, 54–55 *inv*
models of, 4 *smp*
types of, 33–35 *lab*
waning, *35, 36*
waning v. waxing, *36*
waxing, *35, 36*
Lunar Reconnaissance Orbiter (LRO), 37

M

Mass, 78, *79*
Mercury, 105
Mercury Surface, Space Eevironment, Geochemistry, and Ranging (MESSENGER), 105
Meteorites, 83, *108*
Meteoroids, *108*
Meteorological seasons, 19
Meteorologists, 19
Meteors, *108*
Microwaves, 96
Midnight sun, 21
Milky Way, 86
Models
of the inner planets, 100–101 *lab*
of planet's interior, 104 *inv*
of the solar system, 101–103 *lab*

Moons, 106–107 *inv,* 107
exploration of, 37
far side of, 32
light from the Sun and, 30 *inv, 31*
near side of, 32
phases of, 33–35 *lab, 36*
revolution of, 31–32 *inv*
rotation of, 31–32 *inv*

N

NASA, 37, 72 *smp,* 96–97, 105
Nebulae, 80, 82, 83
***New Horizons* mission,** 97
Northern hemisphere
equinoxes and, 17
fall and winter and, *16*
solstices and, 17
spring and summer and, *16*
Nuclear fusion, 80

O

Orbit, 11
distance from Sun and, 98–99 *inv,* 100
of planets, 85

P

Penumbra
defined, **48**
lunar eclipses and, 52
shadows and, 49–50 *lab*
solar eclipses and, 51
Phase, 32
Planets
gravity and, 78, 82
models of, 100–101 *lab,* 102–103 *lab,* 104 *inv*
orbits of, 84–85 *lab,* 98–99 *inv*
rules designating status of, 109
in the solar system, 96–97 *inv*
Pluto, 109
Polar night, 21

R

Radio waves, 96
Ratios, 100–101 *lab,* 101–103 *lab*
Relationships, 98–99 *inv*
Review
Lesson 1, 18–90
Lesson 2, 38–40, 110–112
Lesson 3, 58–60
Revolution, 11
Rotation, 11
Rotation axis
defined, **11**
eclipses and, 54–55 *inv*
effect on poles and, 21
seasons and, 16
tilt of, 11, 14

S

Scale models, 100–101 *lab,* 101–103 *lab,* 104 *inv*
Scarps, 105
Science Probe
Eclipses, 41
Gravity in Space?, 73, 90
Objects in Our Solar System, 91, 112
Phases of the Moon, 25, 40
Seasons, 5, 24
Seasons
Earth's orbit and, 15, *16*
Earth's rotation axis and, 17
Earth's tilt and, 15, *16*
meteorological v. astronomical, 19
models of, 4 *smp*
polar differences and, 21
Shadows
defined, 46
eclipses and, 49–50 *lab*
lunar eclipses and, 52–53
solar eclipses and, 51
umbras and penumbras and, 48
Solar eclipses
defined, ***51***
eye safety and, 57
lunar phases and, 54–55 *inv*
shadows and, 48, 49–50 *lab*
total v. partial, 51–52
Solar system
classification of, 93 *enc,* 96
formation of, 80–81 *lab,* 82, 83
observation of, 96
Solstice, 17, *18,* 19. *see also* **Equinox**
Space exploration, 97
Space probes, 97, 105
Spiral galaxies, *87*
Stars
formation of, 80
galaxies and, 86, *87*
STEM Module Project
Patterns in the Sky, 4, 24, 40, 61–68
Wanted: Space Investigator, 72, 90, 112, 113–119
Sun, *11,* 12

T

Technology, 97
Telescopes, 96–97 *inv*
Temperature, 14, 80
The Solar System, 91–112
Tyson, Neil deGrasse, 109

U

Ultraviolet (UV) rays, 57
Umbra
defined, **48**
lunar eclipses and, 52
shadows and, 49–50 *lab*
solar eclipses and, 51

W

Waning phases, ***35,*** *36*
Watters, Tom, 105
Waxing phases, ***35,*** *36*